Lecture Notes in Mathematics

Edited by A. Dold and B. Eckmann

773

Numerical Analysis

Proceedings of the 8th Biennial Conference
Held at Dundee, Scotland, June 26–29, 1979

Edited by G. A. Watson

Springer-Verlag
Berlin Heidelberg New York 1980

Editor

G. Alistair Watson
University of Dundee
Department of Mathematics
Dundee, DD1 4HN
Scotland

AMS Subject Classifications (1980): 65-06, 65F10, 65F15, 65K05, 65L05, 65N10, 65N30, 65N35

ISBN 3-540-09740-6 Springer-Verlag Berlin Heidelberg New York
ISBN 0-387-09740-6 Springer-Verlag New York Heidelberg Berlin

Library of Congress Cataloging in Publication Data
Main entry under title:
Numerical analysis.
(Lecture notes in mathematics; v. 773)
Bibliography: p.
Includes index.
1. Numerical analysis--Congresses. I. Watson, G. A. II. Series: Lecture notes in mathematics (Berlin); 773.
QA3.L28 no. 773 [QA297] 510s [519.4] 79-28631
ISBN 0-387-09740-6

Printing and binding: Beltz Offsetdruck, Hemsbach/Bergstr.
2141/3140-543210

Preface

For the four days 26 - 29 June, 1979, around 230 people from 26 countries attended the 8th Dundee Biennial Conference on Numerical Analysis at the University of Dundee, Scotland. Invitations to give talks at the meeting were accepted by 13 prominent numerical analysts, representative of a wide variety of fields of activity, and their papers appear in these notes. In addition to the invited papers, short contributions were solicited, and 66 of these, given in three parallel sessions, were presented at the conference. A complete list of these submitted papers, together with authors' addresses, is also given here.

I would like to take this opportunity of thanking the speakers, including the invited after dinner speaker at the conference dinner, Professor J Crank, all chairmen and participants for their contributions. I would also like to thank the many people in the Department of Mathematics of this University who assisted in various ways with the preparation for, and running of, this conference. In particular, I am once more indebted to Mrs R Hume for attending to the considerable task of typing the various documents associated with the conference, and some of the typing in this volume.

Financial support for this conference was obtained from the European Research Office of the United States Army. This support is gratefully acknowledged.

G A Watson

Dundee, November 1979.

CONTENTS

INVITED SPEAKERS

O Axelsson:	Department of Mathematics, Catholic University, Nijmegen, The Netherlands.
J C Butcher:	Department of Mathematics, University of Auckland, Auckland, New Zealand.
E W Cheney:	Department of Mathematics, RLM 8-100, The University of Texas at Austin, Austin, Texas 78712, USA.
L Collatz:	Institut für Angewandte Mathematik, Universitat Hamburg, 2 Hamburg 13, Bundesstr 55, West Germany.
J Crank:	School of Mathematical Studies, Brunel University, Kingston Lane, Uxbridge, Middlesex UB8 3PH, England.
J Cullum:	IBM Thomas J Watson Research Center, Yorktown Heights, New York 10598, USA.
J D Lambert:	Department of Mathematics, University of Dundee, Dundee, Scotland.
J W Jerome:	Department of Mathematics, Northwestern University, Evanston, Illinois 60201, USA.
D Q Mayne:	Department of Computing and Control, Imperial College, London SW7, England.
K W Morton:	Department of Mathematics, University of Reading, Whiteknights, Reading, England.
S P Nørsett:	Department of Mathematics, Institute for Numerical Analysis, N.T.H., N-7034 Trondheim, Norway.
H J Stetter:	Institut für Numerische Mathematik, Technische Hochschule Wien, A-1040 Wien, Gusshausstr, 27-29, Austria.
E L Wachspress:	General Electric Company, Schenectady, New York, USA.
P Wesseling:	Delft University of Technology, Julianalaan 132, Delft, The Netherlands.

SUBMITTED PAPERS

J Albrecht: Institute for Mathematics, Technical University of Clausthal, Germany.
Inclusion theorems for eigenvalues.

C Andrade: Department of Mathematics, University of Sao Paulo, Brazil and
S McKee: The Computing Laboratory, University of Oxford, England.
On optimal high accuracy linear multistep methods for first kind Volterra integral equations.

C T H Baker: Mathematics Department, University of Manchester, England.
Structure of recurrence relations.

J W Barrett: Mathematics Department, University of Reading, England.
An optimal finite element method for a non self-adjoint elliptic operator.

K E Barrett: Mathematics Department, Lanchester Polytechnic, England.
Optimal control methods for heat transfer calculation.

R H Bartels and A R Conn: Department of Combinatorics & Optimization, University of Waterloo, Canada.
An exact penalty algorithm for solving the nonlinear ℓ_1 problem.

H Brunner: Mathematics Department, Dalhousie University, Canada.
The variation of constants formula in the numerical analysis of Volterra equations.

T D Bui: Department of Computer Science, Concordia University, Canada.
Some new L-stable methods for stiff differential systems.

S J Byrne and R W H Sargent: Mathematics Department, Imperial College, London, England.
An algorithm for linear complementarity problems using only elementary principal pivots.

T H Clarysse: Department Wiskunde, University of Antwerp, Belgium.
Rational predictor-corrector methods for nonlinear Volterra integral equations of the second kind.

D B Clegg: Mathematics Department, Liverpool Polytechnic, England.
On Newton's method with a class of rational functions for solving polynomial equations.

J Crank: School of Mathematical Studies, Brunel University, England.
Numerical solution of free boundary problems by interchanging dependent and independent variables.

P E M Curtis: National Physical Laboratory, Teddington, England.
The calculation of optimal aircraft trajectories.

F D'Almeida: Mathematics Department, IMAG, Grenoble, France.
Methods for solving the unsymmetric generalized eigenvalue problem with large matrices issued from the French economy models.

A Davey: Mathematics Department, University of Newcastle upon Tyne.
On the numerical solution of stiff boundary value problems.

L M Delves and C Phillips: Department of Computational and Statistical Science, University of Liverpool, England.
The Global element method - a progress report.

J de Pillis: Mathematics Department, University of California, USA and
M Neumann: Mathematics Department, University of Nottingham, England.
The acceleration of iterative methods via k-part splittings.

W Dickmeis: Rheinisch-Westfälische Technische Hochschule, Aachen, W Germany.
On the Lax-Type equivalence theorems with orders.

S Ellacott: Mathematics Department, Brighton Polytechnic, England.
Numerical conformal mapping - Why bother?

G H Elliott: Mathematics Department, Portsmouth Polytechnic, England.
Economisation in the complex plane.

R Fletcher: Mathematics Department, University of Dundee, Scotland.
An exact L_1 penalty function method for nonlinear equations and nonlinear programming.

H Foerster: G.M.D., St Augustin, W Germany.
Reduction methods for the fast solving of linear elliptic equations.

W Gander: NEU-Tecknikum, Switzerland.
Least squares with a quadratic constraint.

C R Gane: Central Electricity Research Laboratories, Leatherhead, England.
A R Gourlay: IBM United Kingdom Scientific Centre, Winchester, England and
J Ll Morris: Department of Applied Analysis and Computer Science, University of Waterloo, Canada.
From Humble beginnings .....

J-L Gout: Faculty of Science, University of Pau, France.
On a Hermite rational 3th degree finite element.

M H Gutknecht: Mathematics Department, ETH Zurich, Switzerland.
Fast methods to solve Theodorsen's integral equation for conformal mappings.

R J Hanson and K H Haskell: Sandia Laboratories, Albuquerque, USA.
Constrained least squares curve fitting to discrete data using B-splines.

P J Hartley: Mathematics Department, Lanchester Polytechnic, England.
On using curved knot lines.

W D Hoskins: Department of Computer Science, University of Manitoba, Canada and
D J Walton: Department of Mathematical Sciences, Lakehead University, Ontario, Canada.
Improved fourth order methods for the solution of matrix equations of the form $XA + AY = F$.

A Iserles: Department of Applied Mathematics and Theoretical Physics, University of Cambridge, England.
Quadrature method for the numerical solution of O.D.E.

B Kågström: Institute of Information Processing, University of Umeå, Sweden.
How to compute the Jordan normal form - the choice between similarity transformations and methods using the chain relations.

R Kersner: Computer and Automation Institute of Hungarian Academy of Sciences, Budapest, Hungary.
On the properties of solutions of the nonsteady filtration equations with absorption.

D Kraft: Institut für Dynamik der Flugsysteme der DFVLR, Oberpfaffenhofen, West Germany.
Comparing mathematical programming algorithms based on Lagrangian functions for computing optimal aircraft trajectories.

D P Laurie: CSIR, Pretoria, South Africa.
Automatic numerical integration over a triangle.

A V Levy and A C Segura: Universidad Nacional Autónoma de Mexico, Mexico.
Stabilization of Newton's method for the solution of a system of nonlinear equations.

P Lindström: Institute of Information Processing, University of Umeå, Sweden.
A working algorithm based on the Gauss-Newton method for nonlinear least squares problems with nonlinear constraints.

T Lyche: Institute of Informatics, University of Oslo, Norway.
A Newton form for trigonometric Hermite interpolation.

M Mäkelä: Mathematics Department, University of Helsinki, Finland.
On some nonlinear modifications of linear multistep methods.

J C Mason: Mathematics Branch, Royal Military College of Science, Shrivenham, England.
The vector Chebyshev Tau method - A new fast method for simple partial differential equations.

R M M Mattheij: Mathematical Institute, Katholieke Universiteit, Nijmegen, Holland.
A stable method for linear boundary value problem.

S F McCormick: Mathematics Department, Colorado State University, USA.
Mesh refinement methods for $Ax = \lambda Bx$.

J V Miller: Mathematics Department, University of Reading, England.
Adaptive meshes in free and moving boundary problems.

R N Mohapatra: Mathematics Department, American University of Beirut, Lebanon.
Order and class of saturation for some linear operators.

G Moore and A Spence: School of Mathematics, University of Bath, England.
The computation of nontrivial bifurcation points.

N Munksgaard: CE-DATA, Denmark.
Solving sparse symmetric sets of linear equations by preconditioned conjugate gradients.

S Nakazawa: Department of Chemical Engineering, University College of Swansea, Wales.
A note on finite element approximations of convection-diffusion equations.

M R O'Donohoe: Computer Laboratory, Cambridge University, England.
An automatic variable-transformation quadrature scheme for singular integrals.

G Oluremi Olaofe: Mathematics Department, Ibadan, Nigeria.
Quadrature solution of the double eigenvalue problem.

T N Robertson: Mathematics Department, Occidental College, Los Angeles, USA.
Gaussian quadrature applied to Cauchy principal value integrals.

Y Saad: Applied Mathematics Information, University of Grenoble, France.
The method of Arnoldi for computing eigenelements of large unsymmetric matrices.

J M Sanz-Serna: Mathematics Department, University of Valladolid, Spain.
Some aspects of the boundary locus method.

K Schaumberg and J Wasniewski: Mathematics Department, University of Copenhagen and Z Zlatev: The Royal University for Veterinary and Agriculture, Copenhagen, Denmark.
Some results obtained in the numerical solution of oscillatory linear systems of ODE's arising from a chemical problem.

A H Sherman: Department of Computer Science, University of Texas at Austin, USA.
Practical experience with a multi-level method for finite element equations.

A Sidi: Department of Computer Science, Israel Institute of Technology, Israel.
A unified approach to the numerical treatment of integrals with end-point singularities.

S T Sigurdsson: Faculty of Engineering and Science, University of Iceland, Iceland.
A second look at Nørsetts modification of the Adams method.

R B Simpson: Mathematics Department, University of Waterloo, Canada.
A finite element mesh verification algorithm.

S Skelboe: Danish Research Centre for Applied Electronics, Denmark.
Backward differentiation formulas with extended regions of absolute stability.

K Sørli: Institute for Numerical Mathematics, NTH Trondheim, Norway.
An analysis of some explicit alternating direction methods for the numerical solution of the diffusion equation.

G S Stelling: Delft Hydraulics Laboratory, Holland.
Frequency and damping errors of ODE solvers.

P G Thomsen: Institute for Numerical Analysis, The Technical University of Denmark, Denmark.
Jump discontinuities in initial value problems for ordinary differential equations.

R von Seggern: Central Institute for Angewandte Mathematics, KFA, Jülich, West Germany.
Superconvergence by application of the finite element method to linear integro-differential equations.

P-A Wedin: Institute of Information Processing, University of Umeå, Sweden.
Theoretical convergence behaviour of the Gauss-Newton method for nonlinear least squares problems with nonlinear constraints.

K Witsch: G.M.D.-I.M.A., St Augustin, West Germany.
On the condition of discrete boundary value problems.

P H M Wolkenfelt: Mathematical Centre, Amsterdam, Holland.
Stability analysis of numerical methods for second kind Volterra equations by imbedding techniques.

R S Womersley: Mathematics Department, University of Dundee, Scotland.
Uses of a minimax model for nondifferentiable functions.

Y S Wong: Department of Computer Science, University of British Columbia, Canada.
Preconditioned conjugate gradient methods for biharmonic problems.

A GENERALIZED CONJUGATE DIRECTION METHOD AND ITS APPLICATION ON A SINGULAR PERTURBATION PROBLEM

O. Axelsson

ABSTRACT

A generalization of the classical conjugate gradient method to nonsymmetric matrix problems is described. The algorithm has a quasioptimal rate of convergence over the corresponding Krylov set. The application of the algorithm on a class of discretizations of singularly perturbed equations is discussed.

1. INTRODUCTION

We shall consider the calculation of a solution $\hat{\underset{\sim}{u}}$ in a real Hilbert space V, of the linear equation $A\underset{\sim}{u} = \underset{\sim}{a}$, where $A : V \to V$, $\underset{\sim}{a} \in V$, by conjugate gradient type methods. The classical conjugate gradient method is based on the minimization of a functional

$$f(\underset{\sim}{u}) = \tfrac{1}{2}a(\underset{\sim}{u},\underset{\sim}{u})-(\underset{\sim}{c},\underset{\sim}{u})$$

over $\underset{\sim}{u} \in \underset{\sim}{u}^0 \oplus S_k$ at each step k, where $\underset{\sim}{u}^0$ is an arbitrary initial approximation and

$$S_k = \mathrm{SPAN}\{\underset{\sim}{v}^0, B\underset{\sim}{v}^0, \ldots, B^k\underset{\sim}{v}^0\}$$

is the Krylov set (see for instance [9], [10]). We observe that the Krylov set is completely specified by $\underset{\sim}{v}^0$ and B.

The most common choice of $\underset{\sim}{v}^0$ is $\underset{\sim}{v}^0 = \underset{\sim}{r}^0 = B\underset{\sim}{u}^0-\underset{\sim}{b}$, the residual and we shall only consider this choice. B is equal to A or A^TA or may be derived from A by a properly chosen operator C, frequently called preconditioning operator. Further $\underset{\sim}{b} = C\underset{\sim}{a}$. In practice B and $\underset{\sim}{b}$ are not calculated explicitely.

The bilinear form

$$a(\underset{\sim}{u},\underset{\sim}{v}) = (M\underset{\sim}{u},\underset{\sim}{v}),$$

is defined by an inner product in V, where M is self-adjoint and positive definite and $\underset{\sim}{c} = MA^{-1}\underset{\sim}{a}$. In the classical conjugate gradient method, $M = B = A$, if A is symmetric and positive definite over V and $\underset{\sim}{b} = \underset{\sim}{a}$. In the preconditioned version, where both A and C are symmetric and positive definite over V, then $B = CA$ and $M = B^TC^{-1}B$. In the minimal residual conjugate gradient method, $B = M = (CA)^TCA$ and $\underset{\sim}{b} = (CA)^TC\underset{\sim}{a}$.

Due to the minimization property,

$$a(\underset{\sim}{e}^{k+1},\underset{\sim}{v}) = 0, \quad \forall \underset{\sim}{v} \in S_k,$$

where $\underset{\sim}{e}^{k+1} = \underset{\sim}{u}^{k+1} - \hat{\underset{\sim}{u}}$, and we have then

$$a(\underset{\sim}{d}^j, \underset{\sim}{d}^k) = 0, \; j \neq k,$$

where with $\underset{\sim}{r}^k = B\underset{\sim}{u}^k - \underset{\sim}{b}$,

$$\underset{\sim}{d}^k = -\underset{\sim}{r}^k + \beta_{k-1}\underset{\sim}{d}^{k-1},$$

$$\underset{\sim}{u}^{k+1} = \underset{\sim}{u}^k + \lambda_k \underset{\sim}{d}^k, \quad k = 0,1,\ldots, \beta_{-1} = 0.$$

Here $\{\underset{\sim}{d}^k\}$, $\{\underset{\sim}{u}^k\}$ are successive search directions and approximations of $\hat{\underset{\sim}{u}}$, respectively. Formulas for β_k and λ_k are given in Section 2.

Although, in a finite dimensional space R^N, the method apparently is terminating after at most N steps, it is well known that it should be considered as an iterative method in the presence of round-off errors and because the iteration error $\underset{\sim}{e}^k$ is often small enough, even when $k \ll N$. The rate of convergence is determined by the distribution of the Fourier coefficients of $\underset{\sim}{v}^0$ in its expansion in eigenfunctions of B. An upper bound of the number of iterations to reach a relative error $\|\underset{\sim}{e}^k\| / \|\underset{\sim}{e}^0\| \leq \delta'$ is easily found to be

$$k = \mathrm{int}[\ln(\frac{1}{\delta} + \sqrt{\frac{1}{\delta^2} - 1})/\ln\sigma + 1], \tag{1.1}$$

where $\delta = \sqrt{\kappa(M)}\,\delta'$, $\sigma = (1+\kappa^{-1}(B))/(1-\kappa^{-1}(B))$, and where κ is the spectral condition number.

The above is apparently a Ritz method. If we have a nonsymmetric (but coercive and bounded) bilinear form $a(\underset{\sim}{u}, \underset{\sim}{v})$, we may apply the more general Galerkin method instead. This is done in Section 2, where we derive the corresponding conjugate direction method. The name conjugate gradient in this context would perhaps be misleading, since there does not exist a functional and hence no gradient.

We prove that in general we do not have the conjugate orthogonality property $a(\underset{\sim}{d}^j, \underset{\sim}{d}^k) = 0$, $j \neq k$, but only for $j < k$. In this case, all previous search directions $\underset{\sim}{d}^j$ have to be kept along. However, we anticipate that the main use of the new algorithm will be in problems, where few iterations are needed, as for instance, in cases where there can be found a good preconditioning matrix C, such that B = CA has a favourable distribution of eigenvalues (see Section 3).

As for all Galerkin methods, based on a coercive and bounded bilinear form (with boundedness constants $0 < \rho \leq K$) we have a quasioptimal rate of convergence.

In particular, as we shall see this means that for matrices B, which are similarly equivalent to a symmetric and positive definite matrix, an upper bound of the number of iterations are given by (1.1) with $\delta := \delta'\rho/(K\kappa(Q))$, where $\kappa(Q)$ is the spectral condition number of a matrix Q, which transforms B to symmetric form.

In the final section we discuss the application of the algorithm on discretized singularly perturbed equations.

The method presented in this paper is an extension of the method in [7] and [11] and also an extension of a similar method, the modified minimal residual method [4].

2. A CONJUGATE DIRECTION METHOD BASED ON A GALERKIN METHOD

Let $a(u,v)$ be a bilinear form, defined on $V \times V$, where V is a real Hilbert space. We assume that a is coercive,

$$a(u,v) \geq \rho \|u\|^2, \quad \forall u \in V,\ \rho > 0$$

and bounded

$$a(u,v) \leq K\|u\|\,\|v\|, \quad \forall u,v \in V.$$

Here $\|u\| = (u,u)^{\frac{1}{2}}$, where $(\cdot,\cdot)$ is an inner product on V. In some cases, this inner product is defined by a given positive definite selfadjoint operator on V, like $(u,v) = u^T Mv$. In general, in the theoretical analysis we do not explicitly define the inner product, however. We shall consider two bilinear forms

$$a_1(u,v) = (Bu,v), \tag{2.1}$$

where B is a given operator on $V \to V$ and also, to some extent,

$$a_2(u,v) = (Bu,Bv). \tag{2.2}$$

We assume that the symmetric part of B is positive definite. a_2 is symmetric, $a_2(u,v) = a_2(v,u)$, but in general a_1 is obviously not symmetric. However, in the following cases it is symmetric:

(i) B is selfadjoint and commutes with M,

(ii) $B = CA$, C is selfadjoint and positive definite, $M = C^{-1}$ and A is selfadjoint.

We consider the numerical solution of $Bu = b$, $u,b \in V$. This system may have been derived from a given system by preconditioning. For instance, if $Au = a$, $u \in V$, $a \in W$, where A is an operator on $V \to W$ and C on $W \to V$, we let $B = CA$. C is called a preconditioning operator of A. Usually we choose C so that the spectrum of B is more clustered than that of A (see Section 3).

We let $\{d^j\}_{j=0,1,\ldots}$ be search directions and $\{u^j\}$ successive approximations of a solution $\hat{u} \in V$. They are determined by the following recursion formulas,

$$d^k = -r^k + \beta_{k-1} d^{k-1} \tag{2.3}$$

$$u^{k+1} = u^k + \sum_0^k \lambda_j^{(k)} d^j, \quad k = 0,1,2,\ldots,\ \beta_{-1} = 0, \tag{2.4}$$

where

$$r^k = Be^k = Bu^k - b \tag{2.5}$$

is the residual and $e^k = u^k - \hat{u}$ is the error. $\{\lambda_j^{(k)}\}_{j=0}^k$ and β_k, $k = 0,1,\ldots$ are parameters to be determined below.

We have by (2.4)

$$e^{k+1} = e^k + \sum_{j=0}^k \lambda_j^{(k)} d^j \tag{2.6}$$

and by (2.4),(2.5),

$$r^{k+1} = r^k + \sum_{j=0}^k \lambda_j^{(k)} Bd^j. \tag{2.7}$$

Let

$$S_k = \mathrm{SPAN}\{r^0, Br^0, \ldots, B^k r^0\}.$$

Then we see that, by (2.3)

$$\underset{\sim}{d}^k \in S_k,\ \underset{\sim}{e}^k-\underset{\sim}{e}^0 \in S_{k-1} \text{ and } \underset{\sim}{r}^k-\underset{\sim}{r}^0 \in BS_{k-1},$$

so $\underset{\sim}{r}^k \in S_k$.

We determine $\{\lambda_j^{(k)}\}_0^k$ by the Galerkin method

$$a(\underset{\sim}{e}^{k+1},\underset{\sim}{v}) = 0, \quad \forall \underset{\sim}{v} \in S_k. \tag{2.8}$$

Let $\underset{\sim}{v} = \underset{\sim}{d}^\ell$, $\ell = k,k-1,\dots,0$. Then we get from (2.6), (2.8)

$$\sum_{j=0}^{k} \lambda_j^{(k)} a(\underset{\sim}{d}^j,\underset{\sim}{d}^\ell) = -a(\underset{\sim}{e}^k,\underset{\sim}{d}^\ell),\ \ell = k,k-1,\dots,0. \tag{2.9}$$

If $\underset{\sim}{d}^0,\underset{\sim}{d}^1,\dots,\underset{\sim}{d}^k$ are linearly independent, this system of (k+1) linear equations has exactly one solution. This follows since the associated matrix $\Lambda^{(k)}$, defined by

$$\Lambda_{\ell,j}^{(k)} = a(\underset{\sim}{d}^j,\underset{\sim}{d}^\ell),\ 0 \le j,\ \ell \le k, \tag{2.10}$$

satisfies

$$\underset{\sim}{\alpha}^T \Lambda^{(k)} \underset{\sim}{\alpha} = a(\underset{\sim}{u},\underset{\sim}{u}) \ge \rho \|\underset{\sim}{u}\|^2 > 0,\ \underset{\sim}{u} = \sum_{j=0}^{k} \alpha_i \underset{\sim}{d}^j,\quad \forall \underset{\sim}{\alpha} \in R^{k+1},\ \rho > 0,$$

if and only if $\|\underset{\sim}{u}\| \neq 0$, that is $\underset{\sim}{u} \neq \underset{\sim}{0}$. Hence $\Lambda^{(k)}$ is positive definite (that is, the symmetric part of $\Lambda^{(k)}$ is positive definite). By Cramers rule,

$$\lambda_k^{(k)} = -\frac{\det(\Lambda^{(k-1)})}{\det(\Lambda^{(k)})} a(\underset{\sim}{e}^k,\underset{\sim}{d}^k). \tag{2.11}$$

We shall discuss two choices of β_ℓ:

(A) $\quad \beta_\ell = \dfrac{a(\underset{\sim}{d}^\ell,\underset{\sim}{r}^{\ell+1})}{a(\underset{\sim}{d}^\ell,\underset{\sim}{d}^\ell)},\quad \ell = 0,1,\dots,\beta_{-1} = 0$

(B) $\quad \beta_\ell = 0,\quad \ell = -1,0,1,\dots$

The bilinear form a_1.

We consider at first the bilinear form a_1 in (2.1). Then by (2.5), (2.8)

$$(\underset{\sim}{r}^{k+1},\underset{\sim}{v}) = 0,\quad \forall \underset{\sim}{v} \in S_k. \tag{2.12}$$

Let $0 \le j \le \ell-2$. Then

$$a_1(\underset{\sim}{d}^j,\underset{\sim}{r}^\ell) = (B\underset{\sim}{d}^j,\underset{\sim}{r}^\ell) = (\underset{\sim}{r}^\ell,B\underset{\sim}{d}^j) = a_1(\underset{\sim}{e}^\ell,B\underset{\sim}{d}^j) = 0,$$

since $B\underset{\sim}{d}^j \in S_{j+1} \subset S_{\ell-1}$. Hence

$$a_1(\underset{\sim}{d}^j,\underset{\sim}{d}^\ell) = a_1(\underset{\sim}{d}^j,-\underset{\sim}{r}^\ell+\beta_{\ell-1}\underset{\sim}{d}^{\ell-1}) = \beta_{\ell-1}a_1(\underset{\sim}{d}^j,\underset{\sim}{d}^{\ell-1}),$$

and by induction

$$a_1(\underset{\sim}{d}^j,\underset{\sim}{d}^\ell) = \prod_{m=j+1}^{\ell-1} \beta_m a_1(\underset{\sim}{d}^j,\underset{\sim}{d}^{j+1}),\ 0 \le j \le \ell-2. \tag{2.13}$$

For $j = \ell-1$ we have

$$a_1(\underset{\sim}{d}^{\ell-1},\underset{\sim}{d}^\ell) = -a_1(\underset{\sim}{d}^{\ell-1},\underset{\sim}{r}^\ell)+\beta_{\ell-1}a_1(\underset{\sim}{d}^{\ell-1},\underset{\sim}{d}^{\ell-1}).$$

Choice (A):

In choice (A) above, we have $a_1(\underset{\sim}{d}^{\ell-1},\underset{\sim}{d}^{\ell}) = 0$, and hence by (2.13),

$$a_1(\underset{\sim}{d}^{j},\underset{\sim}{d}^{\ell}) = 0,\quad 0 \le j \le \ell-1, \tag{2.14}$$

so $\Lambda^{(k)}$ is then lower triangular. Hence by (2.11), (2.3) and (2.8), we get

$$\lambda_k^{(k)} = -\frac{1}{a_1(\underset{\sim}{d}^k,\underset{\sim}{d}^k)}\, a_1(\underset{\sim}{e}^k,\underset{\sim}{d}^k) = \frac{a_1(\underset{\sim}{e}^k,\underset{\sim}{r}^k)}{a_1(\underset{\sim}{d}^k,\underset{\sim}{d}^k)},$$

i.e.

$$\lambda_k^{(k)} = \frac{(\underset{\sim}{r}^k,\underset{\sim}{r}^k)}{a_1(\underset{\sim}{d}^k,\underset{\sim}{d}^k)}. \tag{2.15}$$

Hence $\lambda_k^{(k)} > 0$, unless $\underset{\sim}{r}^k = 0$ (that is unless we already have a solution), in which case $\lambda_k^{(k)} = 0$. Furthermore, (2.9) has a unique solution as long as $a(\underset{\sim}{d}^{(\ell)},\underset{\sim}{d}^{(\ell)}) = 0$, $\ell = 0,1,\dots,k$, that is, as long as $\underset{\sim}{d}^{(\ell)} \neq \underset{\sim}{0}$. When $\lambda_\ell^{(\ell)} > 0$, by (2.7), (2.12), (2.14), the formula for β_ℓ can be simplified,

$$\beta_\ell = \frac{(B\underset{\sim}{d}^\ell,\underset{\sim}{r}^{\ell+1})}{(B\underset{\sim}{d}^\ell,\underset{\sim}{d}^\ell)} = \frac{(\underset{\sim}{r}^{\ell+1}-\underset{\sim}{r}^\ell,\underset{\sim}{r}^{\ell+1})}{(\underset{\sim}{r}^{\ell+1}-\underset{\sim}{r}^\ell,\underset{\sim}{d}^\ell)}.$$

Hence in case $a = a_1$, we have by (2.3), (2.12)

$$\beta_\ell = \frac{(\underset{\sim}{r}^{\ell+1},\underset{\sim}{r}^{\ell+1})}{(\underset{\sim}{r}^\ell,\underset{\sim}{r}^\ell)} = \left(\frac{\|\underset{\sim}{r}^{\ell+1}\|}{\|\underset{\sim}{r}^\ell\|}\right)^2 \tag{2.16}$$

The corresponding method we call the generalized conjugate direction method. If a_1 is a symmetric bilinear form, then $\Lambda^{(k)}$ is a symmetric matrix, and hence with this choice of β_ℓ, $\Lambda^{(k)}$ is diagonal. We then recognize the classical conjugate gradient method (see for instance [9]). The property $a_1(\underset{\sim}{d}^j,\underset{\sim}{d}^\ell) = 0$, $0 \le j \le \ell-1$, corresponds to the conjugate orthogonality among the search directions in that latter method, because when a_1 is symmetric, $a_1(\underset{\sim}{d}^j,\underset{\sim}{d}^\ell) = 0$, all $j \neq \ell$.

Case (B):

In case (B), where $\beta_\ell = 0$, we have by (2.3), (2.13),

$$a_1(\underset{\sim}{d}^{j},\underset{\sim}{d}^{\ell}) = 0,\quad 0 \le j \le \ell-2, \tag{2.17}$$

and $\Lambda^{(k)}$ is a lower Hessenberg matrix.

Consider now the special case where $B = I - M^{-1}N$, where M is symmetric and positive definite and N is skew symmetric. (This means, that the preconditioning operator C, introduced in section 1, is $C = M = \frac{1}{2}(A+A^T)$, the symmetric part of A and $N = \frac{1}{2}(A^T - A)$. We naturally have to assume that the symmetric part is positive definite.) Then,

$$-a_1(\underset{\sim}{d}^j,\underset{\sim}{d}^\ell) = a_1(\underset{\sim}{r}^j,\underset{\sim}{d}^\ell) = (B\underset{\sim}{r}^j,\underset{\sim}{d}^\ell) = \underset{\sim}{r}^{j^T}(M+N)d^\ell = (\underset{\sim}{r}^j,(I+M^{-1}N)\underset{\sim}{d}^\ell).$$

Since apparently $M^{-1}N\underset{\sim}{d}^\ell = (I-B)\underset{\sim}{d}^\ell \in S_{\ell+1}$, we have by (2.12),

$$(B\underset{\sim}{r}^j,\underset{\sim}{d}^\ell) = -a_1(\underset{\sim}{d}^j,\underset{\sim}{d}^\ell) = 0,\quad \ell \le j-2. \tag{2.18}$$

Hence $a_1(\underset{\sim}{d}^j,\underset{\sim}{d}^\ell) = 0$, $j \neq \ell-1,\ell,\ell+1$, so by (2.17) $\Lambda^{(k)}$ is tridiagonal. Since $\lambda_j^{(j)} \neq 0$, we see that we can write (2.7) as

$$\underset{\sim}{r}^{k+1} = \underset{\sim}{r}^k-\lambda_k B\underset{\sim}{r}^k + \sum_{j=0}^{k} \gamma_j^{(k)}\underset{\sim}{r}^j. \tag{2.19}$$

By (2.12), (2.18) we get then

$$(\underset{\sim}{r}^{k+1},\underset{\sim}{r}^j) = 0 \Rightarrow \gamma_j^{(k)}(\underset{\sim}{r}^j,\underset{\sim}{r}^j) = 0, \quad 0 \le j \le k-2,$$

so $\gamma_j^{(k)} = 0$, $0 \le j \le k-2$. Further,

$$(\underset{\sim}{r}^{k+1},\underset{\sim}{r}^k) = 0 \Rightarrow \lambda_k(B\underset{\sim}{r}^k,\underset{\sim}{r}^k)-\gamma_k^{(k)}(\underset{\sim}{r}^k,\underset{\sim}{r}^k) = (\underset{\sim}{r}^k,\underset{\sim}{r}^k).$$

Since N is skewsymmetric, we have

$$(B\underset{\sim}{r}^k,\underset{\sim}{r}^k) = \underset{\sim}{r}^{k^T}(M+N)\underset{\sim}{r}^k = \underset{\sim}{r}^{k^T}M\underset{\sim}{r}^k = (\underset{\sim}{r}^k,\underset{\sim}{r}^k).$$

Hence

$$\gamma_k^{(k)} = \lambda_k-1. \tag{2.20}$$

Further, by (2.12)

$$(\underset{\sim}{r}^{k+1},\underset{\sim}{r}^{k-1}) = 0 \Rightarrow \lambda_k(B\underset{\sim}{r}^k,\underset{\sim}{r}^{k-1})-\gamma_{k-1}^{(k)}(\underset{\sim}{r}^{k-1},\underset{\sim}{r}^{k-1}) = 0. \tag{2.21}$$

Since

$$(B\underset{\sim}{r}^k,\underset{\sim}{r}^{k-1}) = \underset{\sim}{r}^{k^T}(M+N)\underset{\sim}{r}^{k-1} = \underset{\sim}{r}^{k^T}M(2I-B)\underset{\sim}{r}^{k-1} = (\underset{\sim}{r}^k,-B\underset{\sim}{r}^{k-1}),$$

we have by (2.19), (2.12),

$$(B\underset{\sim}{r}^k,\underset{\sim}{r}^{k-1}) = \frac{1}{\lambda_{k-1}}(\underset{\sim}{r}^k,\underset{\sim}{r}^k).$$

Hence, by (2.21)

$$\lambda_{k-1}^{(k)} = \frac{\lambda_k}{\lambda_{k-1}}\,\frac{(\underset{\sim}{r}^k,\underset{\sim}{r}^k)}{(\underset{\sim}{r}^{k-1},\underset{\sim}{r}^{k-1})}. \tag{2.22}$$

From (2.20), (2.22) we see that among the three parameters λ_k, $\gamma_k^{(k)}$, $\gamma_{k-1}^{(k)}$, we can choose one parameter freely except that $\lambda_k \neq 0$. We choose $\gamma_{k-1}^{(k)} = -\gamma_k^{(k)}$. Then we get

$$\lambda_k^{-1} = 1 + \lambda_{k-1}^{-1}\frac{(\underset{\sim}{r}^k,\underset{\sim}{r}^k)}{(\underset{\sim}{r}^{k-1},\underset{\sim}{r}^{k-1})}, \quad k = 1,2,\dots, \quad \lambda_0 = 1$$

and (2.19) simplifies to

$$\underset{\sim}{r}^{k+1} = \underset{\sim}{r}^{k-1} - \lambda_k(B\underset{\sim}{r}^k-\underset{\sim}{r}^k+\underset{\sim}{r}^{k-1}), \quad k = 1,2,\dots\,. \tag{2.23}$$

The corresponding recursion formula for $\underset{\sim}{u}^k$ is

$$\underset{\sim}{u}^{k+1} = \underset{\sim}{u}^{k-1} - \lambda_k(\underset{\sim}{r}^k-\underset{\sim}{u}^k+\underset{\sim}{u}^{k-1}), \quad k = 1,2,\dots,$$

$\underset{\sim}{u}^0$ arbitrarily.

This is the "generalized conjugate gradient method", derived in another way in [7] and [11].

The bilinear form a_2.

Consider now the case

$$a(\underset{\sim}{u},\underset{\sim}{v}) = (B\underset{\sim}{u},V\underset{\sim}{u}).$$

Then, with the choice (A) of β_ℓ we have

$$\beta_\ell = (B\underset{\sim}{r}^{\ell+1}, B\underset{\sim}{d}^\ell) / (B\underset{\sim}{d}^\ell, B\underset{\sim}{d}^\ell).$$

This reduces to the modified minimal residual method, presented in [4]. In this case $\Lambda^{(k)}$ is a symmetric matrix. If B is symmetric it follows from [4] that $\Lambda^{(k)}$ is diagonal, and we have then the minimal residual conjugate gradient method. In the general case, where B is not symmetric, we have to work with all the previous search directions. In practice, however, one finds that the algorithm often works well even if only a few search directions are kept along. This is discussed in [4].

Alternatively, we may solve the normal equations $B^T(B\underset{\sim}{u}-\underset{\sim}{b}) = 0$, where obviously B^TB is symmetric, but only at the expense of an increased condition number and hence an increased number of iterations.

3. QUASIOPTIMAL RATE OF CONVERGENCE

If a is a symmetric form we have, by Ritz principle,

$$f(\underset{\sim}{u}^k) = \min_{\underset{\sim}{v}\in\underset{\sim}{u}^0\oplus S_k} f(\underset{\sim}{v}) = \min_{\underset{\sim}{v}\in\underset{\sim}{u}^0\oplus S_k} \{\tfrac{1}{2}a(\underset{\sim}{v},\underset{\sim}{v}) - (\underset{\sim}{b},\underset{\sim}{v})\},$$

which implies monotonicity,

$$a(\underset{\sim}{e}^{k+1},\underset{\sim}{e}^{k+1}) < a(\underset{\sim}{e}^k,\underset{\sim}{e}^k)$$

if $\underset{\sim}{r}^k \neq 0$, and optimal rate of convergence with respect to the "energy norm" $\{a(\cdot,\cdot)\}^{\frac{1}{2}}$. Since a_1 is not symmetric in general, we cannot prove monotonicity, however. Furthermore, since there does then not exist a functional corresponding to $a(\cdot,\cdot)$, we can only prove a quasioptimal rate of convergence. The proof of quasioptimality is the standard one for Galerkin methods, and is presented here only for completeness. We have

$$a(\hat{\underset{\sim}{u}},\underset{\sim}{v}) = (\underset{\sim}{b},\underset{\sim}{v}) \qquad \forall \underset{\sim}{v} \in V$$

and

$$a(\underset{\sim}{u}^k,\underset{\sim}{v}) = (\underset{\sim}{b},\underset{\sim}{v}) \qquad \forall \underset{\sim}{v} \in S_{k-1} \subset V,$$

so

$$a(\underset{\sim}{e}^k,\underset{\sim}{v}) = 0 \qquad \forall \underset{\sim}{v} \in S_{k-1}.$$

Hence, by the coercivity and boundedness of a,

$$\rho\|\underset{\sim}{e}^k\|^2 \le a(\underset{\sim}{e}^k,\underset{\sim}{e}^k) = a(\underset{\sim}{e}^k,\hat{\underset{\sim}{u}}-\underset{\sim}{v}) \le K\|\underset{\sim}{e}^k\|\,\|\hat{\underset{\sim}{u}}-\underset{\sim}{v}\|, \quad \forall \underset{\sim}{v} \in \underset{\sim}{u}^0 \oplus S_{k-1}.$$

Hence

$$\|\underset{\sim}{e}^k\| \le \frac{K}{\rho} \min_{\underset{\sim}{v}\in u^0 \oplus S_{k-1}} \|\hat{\underset{\sim}{u}}-\underset{\sim}{v}\|. \tag{3.1}$$

In this way, we have reduced the problem of estimating the rate of convergence to that of estimating the best approximation of $\hat{u} \in V$ by $\underset{\sim}{v} \in \underset{\sim}{u}^0 \oplus S_{k-1}$.

As is wellknown, see for instance [3], if B is symmetric and positive definite, the rate of convergence of $\min_{\underset{\sim}{v} \in \underset{\sim}{u}^0 \oplus S_{k-1}} \|\hat{\underset{\sim}{u}}-\underset{\sim}{u}\|$, $k \to \infty$, is determined by the distribution of the eigenvalues of the operator B and of the Fourier coefficients in the development of $\underset{\sim}{e}^0$ in the eigenvectors of B. An upper bound of the number of iterations are given by (1.1).

If B is similarily equivalent to a symmetric matrix with positive eigenvalues, i.e. if there exists a nonsingular matrix Q such that $Q^{-1}BQ = \tilde{B}$ is symmetric and positive definite, a similar bound can be derived. Let π_k^0 be the set of polynomials of degree k that take the value 1 at the origin. We have

$$\min_{\underset{\sim}{v} \in \underset{\sim}{u}^0 \oplus S_{k-1}} \|\hat{\underset{\sim}{u}}-\underset{\sim}{v}\| = \min_{\underset{\sim}{v} \in S_{k-1}} \|\underset{\sim}{e}^0-\underset{\sim}{v}\| .$$

so we get

$$\min_{\underset{\sim}{v} \in \underset{\sim}{u}^0 \oplus S_{k-1}} \|\hat{\underset{\sim}{u}}-\underset{\sim}{v}\| = \min_{p_k \in \pi_k^0} \|p_k(B)\underset{\sim}{e}^0\| \le \|Q\| \min_{p_k \in \pi_k^0} \|p_k(\tilde{B})Q^{-1}\underset{\sim}{e}^0\|$$

$$\le \kappa(Q) \min_{p_k \in \pi_k^0} \max_{\lambda \in S(B)} |p_k(\lambda)| \, \|\underset{\sim}{e}^0\|,$$

where S(B) is the spectrum of B, that is, also of $\tilde{B}$ and where $\kappa(Q) = \|Q\| \, \|Q^{-1}\|$. Hence we see from (3.1) that the relative iteration error satisfies

$$\|\underset{\sim}{e}^k\| / \|\underset{\sim}{e}^0\| \le \frac{K}{\rho} \kappa(Q) \min_{p_k \in \pi_k^0} \max_{\lambda \in S(B)} |p_k(\lambda)| .$$

It is wellknown that

$$\min_{p_k \in \pi_k^0} \max_{\lambda \in S(B)} |p_k(\lambda)| \le T_k\left(\frac{1+\kappa^{-1}(B)}{1-\kappa^{-1}(B)}\right)^{-1},$$

where

$$T_k(z) = \tfrac{1}{2}[(z+(z^2-1)^{\frac{1}{2}})^k + (z-(z^2-1)^{\frac{1}{2}})^k],$$

is the Chebyshev polynomial. Hence, the upper bound of k, such that $\|\underset{\sim}{e}^k\| / \|\underset{\sim}{e}^0\| \le \delta'$, $\delta = \delta'\rho/K\kappa(Q)$, as given in (1.1), follows.

If B = CA, where C and A are symmetric and positive definite, we have

$$\tilde{B} = C^{-\frac{1}{2}}BC^{\frac{1}{2}} = C^{\frac{1}{2}}AC^{\frac{1}{2}},$$

i.e. $Q = C^{\frac{1}{2}}$. Hence we can then expect that $\kappa(Q)$ is not large in practice, so the number of iterations does essentially depend only on the eigenvalues of B and of the Fourier coefficients. To some extent it may also depend on K/ρ, but due to the logarithm function in (1.1), the influence of K/ρ is usually minor.

In the following section we shall however consider an example where $\kappa(Q)$ may grow exponentially with respect to a problem parameter, and then naturally $\kappa(Q)$ influences the number of iterations.

4. APPLICATION ON A SINGULAR PERTURBED PROBLEM

Consider the singular perturbed problem

(4.1) $$Lu = -\varepsilon\Delta u + \vec{v}\nabla u + cu = f, \quad \underset{\sim}{x} = (x_1,x_2,x_3) \in \Omega \subset \mathbb{R}^3,$$

where $c \geq 0$, $\varepsilon > 0$ and u is given on $\partial\Omega$. We assume that $|\vec{v}| \neq 0$, that $v_i(\underset{\sim}{x}) = v_i(x_i)$, $i = 1,2,3$, and that Ω is a rectangular region. Although these assumptions are somewhat restrictive we are still able to get interesting information regarding the generalized conjugate direction method, when applied on this type of problems.

We let (4.1) be discretized by an up wind difference scheme or by a similar positive scheme. We assume that the symmetric part of the associated matrix $\mathcal{A}$ is positive definite, so that

$$a(\underset{\sim}{u},\underset{\sim}{u}) = \underset{\sim}{u}^T \mathcal{A} \underset{\sim}{u} \geq \rho \underset{\sim}{u}^T \underset{\sim}{u}, \quad \rho \geq \varepsilon > 0$$

A sufficient condition for this to be true is that $\mathrm{div}(\vec{v}(x)) \leq 0$, $\underset{\sim}{x} \in \Omega$. This follows from the variational formulation or from the selfadjoint part $\frac{1}{2}(L+L^*)$ of the continuous operator, since

$$\tfrac{1}{2}(L+L^*)u = -\varepsilon\Delta u + (c-\mathrm{div}(\vec{v}))u.$$

The structure of the associated matrix is

$$\mathcal{A} = \begin{bmatrix} (A+D_1) & c_1 I & & \\ b_1 I & (A+D_2) & c_2 I & \\ & \ddots & \ddots & \ddots \\ & & b_{n-1} I & (A+D_n) \end{bmatrix}$$

where D_i are positive diagonal matrices and $b_i, c_i < 0$. A is also block-tridiagonal in a three dimensional problem, and tridiagonal with negative off-diagonal coefficients in a plane problem. All diagonal entries are positive. Hence both A and $\mathcal{A}$ are quasisymmetric, that is, there exists a diagonal matrix D such that DAD^{-1} is symmetric and $\mathcal{D} = \mathrm{diag}(\delta_i D)$, such that

$$\mathcal{D}\mathcal{A}\mathcal{D}^{-1} = \begin{bmatrix} (DAD^{-1}+D_1) & (b_1c_1)^{\frac{1}{2}} I & & \\ (b_1c_1)^{\frac{1}{2}} I & (DAD^{-1}+D_2) & (b_2c_2)^{\frac{1}{2}} I & \\ & \ddots & \ddots & \ddots \\ & & (b_{n-1}c_{n-1})^{\frac{1}{2}} I & (DAD^{-1}+D_n) \end{bmatrix}$$

Here $\delta_{i+1} = (c_i/b_i)^{\frac{1}{2}}\delta_i$, $i = 1,2,\ldots,n-1$, $\delta_0 = 1$.

Hence, with the similarity transformation matrix $Q = \mathcal{D}$, we realize that since $b_i/c_i = O(\varepsilon)$ or $O(\varepsilon^{-1})$,

(4.2) $$\kappa(Q) = O(\varepsilon^{-\frac{n}{2}d}), \ \varepsilon \to 0, \text{ where d is the space dimension.}$$

Since the off-diagonal entries of $\tilde{B}$ are $O(\varepsilon^{\frac{1}{2}})$, the spectrum of $\tilde{B}$ is contained in a circle of radius $O(\varepsilon^{\frac{1}{2}})$ about the point (1,0) in the complex plane. (Without limitation, we may assume that $\mathcal{A}$ is diagonally scaled, so that its diagonal entries are 1.) From this, (1.1) and (4.2), it follows that the upper bound of the number of iterations,

$$k^* = O(\ln \kappa(Q)) = \frac{nd}{2} O(\ln \varepsilon^{-1}), \quad \varepsilon \to 0.$$

Hence, for small enough, but fixed ε, we can expect that the number of iterations grows proportional to n. Since $n = O(h^{-1})$, we have then $k^* = O(h^{-1})$, $h \to 0$. This bound is also valid for regular problems, where $h << \varepsilon$ (see [3]).

The operation cost of each iteration in the application of the generalized conjugate direction method is proportional to ℓN at the ℓ'th iteration step, where $N = O(n^d)$ is the number of unknowns. The number of operations in the solution of (2.9) is only $O(\ell^2)$, since $\Lambda^{(\ell)}$ is a triangular matrix. The operation count in matrix-vector multiplications $B\underset{\sim}{d}^{\ell}$ etc. is $O(N)$ and in the construction of a new column in $\Lambda^{(\ell)}$ is $(\ell+1)N$, that is of not higher order than $O(\ell N)$, $n \to \infty$.

The total number of operations in the generalized conjugate direction method (based on $B = A$) is hence bounded above by $O(n^{d+2})$, $n \to 0$.

Let us make a comparison with the number of operations in a direct solver, based on an LU-decomposition method, assuming a natural ordering of the grid points. Then the LU-factorization demands $O(n^{3d-2})$ operations and the number of operations is less in the iterative solver if $d = 3$, if only h is small enough. Furthermore, numerical tests indicate that the number of iterations (and operations) is lowered considerably if the inverse of an incomplete factorization of A is applied as a preconditioning matrix C (see [5]).

Let us finally compare the number of operations if the corresponding normal equations are solved (that is, if a least square method is used). Then one finds that, in a plane problem the corresponding continuous operator is L^*L, which, assuming $c = 0$ in (4.1) is defined by

$$L^*Lu = \varepsilon^2\Delta^2 u - \varepsilon\left(\frac{\partial v_1}{\partial x_1} - \frac{\partial v_2}{\partial x_2}\right)\left(\frac{\partial^2 u}{\partial x_1^2} - \frac{\partial^2 u}{\partial x_2^2}\right) - 2\varepsilon\left(\frac{\partial v_1}{\partial x_2} + \frac{\partial v_2}{\partial x_1}\right)\frac{\partial^2 u}{\partial x_1 \partial x_2}$$

$$- \varepsilon\Delta(\vec{v}).\nabla u v - \operatorname{div}((\vec{v}.\nabla u)\vec{v}).$$

The reduced equation operator (derived when $\varepsilon = 0$), is however not strongly elliptic, since we may have $\vec{v}.\nabla u \equiv 0$ on a nontrivial subset Ω' of Ω. Hence, in general, the condition number of the discretized equation is $O(n^2+\varepsilon^2 n^4)$ if $\vec{v}.\nabla u \not\equiv 0$ in Ω but may be $O(n^4)$ if $\vec{v}.\nabla u \equiv 0$ in Ω'. Hence, a (classical) conugate gradient (or direction) method would demand $O(n^2)$, $n \to 0$, iterations, if it is not preconditioned. If preconditioned, the number of iterations would however only be $O(n^{3/2})$ (see for instance [1] and [2]), or even $O(n)$ (see [6], [8]), the latter being also the rate of convergence for the unpreconditioned method applied on A, the discretization of L.

For regular problems, that is, problems where $h << \varepsilon$, it has been proven in [11] a superlinear rate of convergence of the generalized conjugate gradient method (preconditioned by the symmetric part).

5. CONCLUSIONS

We have seen that the generalized conjugate direction method to solve unsymmetric problems may be a powerful method if not too many iterations are needed. In particular, the number of operations for a singular perturbation problem, preconditioned by an incomplete LU-factorization method, seems to be of optimal order. For other problems one finds often in practice that one does not have to carry along all old search directions, but only a few of the newest ones. Continued research to give more theoretical results in the application of the new generalized conjugate direction method to more general classes of problems would however be of interest.

REFERENCES

1. Andersson,L., SSOR preconditioning of Toeplitz matrices, Report 76.02R (Ph.D. thesis), Department of Computer Sciences, Chalmers University of Technology, Göteborg, Sweden (1976).
2. Axelsson,O., Notes on the numerical solution of the biharmonic equations, J.Inst.Maths. Applics. 11(1973), 213-226.
3. Axelsson,O., A class of iterative methods for finite element equations, Computer Methods in Applied Mathematics and Engineering 9(1976), 123-137.
4. Axelsson,O., Conjugate gradient type methods for unsymmetric and inconsistent systems of linear equations, to appear Linear Algebra and its Applications.
5. Axelsson,O., and Gustafsson,I., A modified upwind scheme for convective transport equations and the use of a conjugate gradient method for the solution of non-symmetric systems of equations, J.Inst. Maths. Applics. 23(1979), 321-337.
6. Axelsson,O., and Gustafsson,I., An iterative solver for a mixed variable variational formulation of the (first) biharmonic problem, to appear in Comp. Math.Appl.Mech.Eng.
7. Concus,P., and Golub,G.H., A generalized conjugate gradient method for non-symmetric systems of linear equations, Proc.Second Internat.Symp. on Computing Methods in Applied Sciences and Engineering, IRIA (Paris, Dec.1975), Lecture Notes in Economics and Mathematical Systems, vol.134, R.Glowinski and J.-L.Lions, eds., Springer-Verlag, Berlin 1976.
8. Gustafsson,I., Stability and rate of convergence of modified incomplete Cholesky factorization methods, Report 79.02R (Ph.D. thesis), Department of Computer Sciences, Chalmers University of Technology, Göteborg, Sweden (1979).
9. Hestenes,M.R., and Stiefel,E., Method of conjugate gradients for solving linear systems, J.Res.Nat.Bur. Standards, No.49, 409-436, 1952.
10. Lanczos,C., Solutions of the systems of linear equations by minimized operations, J.Res.Nar.Bur. Standards, No.49, 33-53, 1952.
11. Widlund,O., A Lanczos' method for a class of nonsymmetric systems of linear equations, SIAM J.Num.Anal. 15(1978), 801-812.

SOME IMPLEMENTATION SCHEMES FOR IMPLICIT RUNGE-KUTTA METHODS

J. C. Butcher

1. INTRODUCTION

In the panel discussion on the implementation of implicit Runge-Kutta methods at the Numerical Ordinary Differential Equations meeting held in Champaign, Illinois in April 1979, many different approaches were surveyed. In particular, Dr T.A. Bickart presented a detailed comparison of the operation counts and memory requirements for techniques published either formally or informally by himself, by M.A. Epton, by G.J. Tee and by the present author.

One of these approaches requires the use of methods whose coefficient matrix posseses a simple spectrum, preferably containing only a single point. This is a natural generalization of the use of semi-explicit methods as proposed by S.P. Nørsett [11]and by R. Alexander [1], and allows a technique proposed by the author [5] to show itself to greatest effect. In papers by K. Burrage [3] and the author [6], a class of methods possessing the one point spectrum property is described and they and F.H. Chipman have constructed an algorithm known as STRIDE based on these methods [4], [7].

Finally, the iterated defect correction procedure is a promising approach. Although this technique is most naturally described as applying to block implicit methods, which are equivalent to implicit Runge-Kutta methods with an equally spaced internal mesh, it can be generalized to a wider class of implicit Runge-Kutta methods. Papers by R. Frank and C. Überhuber [9], [10] describe this procedure.

The method that will be introduced here, has some formal similarities with the iterated defect correction procedure but is applicable to a much wider class of implicit Runge-Kutta methods. In particular, they need not be collocation methods and there is no reason for the internal points of the method to be equally spaced or, for that matter, distinct.

The new method also has similarities to a proposal of Chipman [8] but in many situations has a better convergence rate. It is applicable more widely than semi-explicit methods, does not require similarity transformations to be added to the computational costs, and appears to have a speed advantage over the method of Bickart [2] because of a lower operation count per iteration.

2. MOTIVATION THROUGH ITERATED DEFECT CORRECTION

In this section we describe the iterated defect correction method from a point of view slightly different from the usual one, and show how it suggests a wider class of iteration procedures.

Let

$$\begin{array}{c|cccc} c_1 & a_{11} & a_{12} & \cdots & a_{1s} \\ c_2 & a_{21} & a_{22} & \cdots & a_{2s} \\ \vdots & \vdots & \vdots & & \vdots \\ c_s & a_{s1} & a_{s2} & \cdots & a_{ss} \\ \hline & b_1 & b_2 & \cdots & b_s \end{array} = \begin{array}{c|c} c & A \\ \hline & b^T \end{array} \tag{2.1}$$

be an implicit Runge-Kutta method where the matrix A is assumed to be non-singular. If y_{n-1} is the computed solution at the beginning of step n for the differential equation $y'(x) = f(y(x))$, and $Y_1,Y_2,\ldots,Y_s$ are the internal approximations computed at the s stages of the method then $Y_1,Y_2,\ldots,Y_s,y_n$ are defined by

$$\begin{aligned} Y_i &= y_{n-1} + h \sum_{j=1}^{s} a_{ij} f(Y_j) \ , \quad i = 1,2,\ldots,s \ , \\ y_n &= y_{n-1} + h \sum_{j=1}^{s} b_j f(Y_j) \ . \end{aligned} \tag{2.2}$$

In the iterated defect correction method, initial approximations to $Y_1,Y_2,\ldots,Y_s$ are found by a semi-explicit method

$$\begin{array}{c|cccc} \bar{c}_1 & \bar{a}_{11} & 0 & \cdots & 0 \\ \bar{c}_2 & \bar{a}_{21} & \bar{a}_{22} & \cdots & 0 \\ \vdots & \vdots & \vdots & & \vdots \\ \bar{c}_s & \bar{a}_{s1} & \bar{a}_{s2} & \cdots & \bar{a}_{ss} \\ \hline \end{array} = \bar{c} \,\Big|\, \bar{A}$$

which we will refer to as the crude method. Denote these initial approximations by $Y_1^0, Y_2^0, \ldots, Y_s^0$ so that

$$(2.3) \qquad Y_i^0 = y_{n-1} + h \sum_{j=1}^{i} \bar{a}_{ij} f(Y_j^0) \quad , \; i = 1,2,\ldots,s.$$

A sequence of further approximations $Y_1^m, Y_2^m, \ldots, Y_s^m$ (m=1,2,...) is defined by

$$(2.4) \qquad Y_i^m = Y_i^0 - (Z_i^{m-1} - Y_i^{m-1})$$

with $Z_1^m, Z_2^m, \ldots, Z_s^m$ (m=0,1,...) computed using a modified form of (2.3) namely

$$(2.5) \qquad Z_i^m = y_{n-1} + h \sum_{j=1}^{i} \bar{a}_{ij} (f(Z_j^m) + d_j^m), \; i = 1,2,\ldots,s,$$

where the so-called defect is defined by

$$d_i^m = (\frac{1}{h} \sum_{j=1}^{s} u_{ij} (Y_j^m - y_{n-1}) - f(Y_i^m)) \quad , \; i = 1,2,\ldots,s,$$

for u_{ij} (i,j=1,2,...,s) elements of some matrix U.

A typical form of $\bar{A}$ is

$$\begin{bmatrix} c_1 & 0 & 0 & \cdots & 0 \\ c_1 & c_2-c_1 & 0 & \cdots & 0 \\ c_1 & c_2-c_1 & c_3-c_2 & \cdots & 0 \\ \vdots & \vdots & \vdots & & \\ c_1 & c_2-c_1 & c_3-c_2 & \cdots & c_s-c_{s-1} \end{bmatrix}$$

so that $\bar{c}_1 = c_1, \bar{c}_2 = c_2, \ldots, \bar{c}_s = c_s$. With this choice, the internal stages of the crude method can be interpreted as s steps of the implicit Euler method using stepsizes $c_1 h$, $(c_2 - c_1)h, \ldots, (c_s - c_{s-1})h$ respectively. With equal spacing, $c_i = i/s$ (i=1,2,...,s), these stages of the crude method can be conveniently implemented each using the triangular factors of the same matrix $(I - \frac{1}{s}hJ)$ for J an approximation to the Jacobian matrix at a recent point on the solution trajectory.

In the case of collocation methods, the matrix U is defined by the numerical differentiation formula

$$p'(x_{n-1}+hc_i) = \frac{1}{h}\sum_{j=1}^{s} u_{ij}(p(x_{n-1}+hc_j) - p(x_{n-1})),$$

$$i = 1,2,\ldots,s$$

which is to hold exactly for p any polynomial of degree not exceeding s. We extend this immediately to non-collocation methods by choosing $U = A^{-1}$ to ensure that the limiting value of the vector $[Y_1^m, Y_2^m, \ldots, Y_s^m]$ as $m \to \infty$, if this limit exists, gives the solution for the original method (2.1).

From (2.3), (2.4), (2.5) we obtain

$$(2.6) \qquad Y_i^m = y_{n-1} + h\sum_{j=1}^{i} \bar{a}_{ij}(f(Y_j^0) - f(Z_j^{m-1}) + f(Y_j^{m-1}))$$

$$+ \sum_{j=1}^{s} v_{ij}(Y_j^{m-1} - y_{n-1}), \quad i = 1,2,\ldots,s,$$

where the matrix V, with elements v_{ij} $(i,j = 1,2,\ldots,s)$, is given by

$$(2.7) \qquad V = I - \bar{A}U = I - \bar{A}A^{-1}.$$

The innovation proposed in this paper is the replacement in (2.6) of $(f(Y_j^0) - f(Z_j^{m-1}) + f(Y_j^{m-1}))$ by $f(Y_j^0 - Z_j^{m-1} + Y_j^{m-1})$ which, by (2.4), will equal $f(Y^m)$. With this change, we obtain an iteration scheme given by

$$(2.8) \qquad Y_i^m = y_{n-1} + h\sum_{j=1}^{i} \bar{a}_{ij} f(Y_j^m) + \sum_{j=1}^{s} v_{ij}(Y_j^{m-1} - y_{n-1}),$$

$$i = 1,2,\ldots,s$$

which, when it converges, does so to the same limit and at a similar rate as for the iterated defect correction scheme. It is, in fact, identical to it in the case of linear differential equation systems.

What is lost, of course, is the close link with the theory of P.E. Zadunaisky but there are gains that compensate for this loss. In particular, it is no longer necessary for iteration number 0 to be produced by (2.3) and $Y_1^0, Y_2^0, \ldots, Y_s^0$ could, for example, be computed as extrapolates from previous steps. Furthermore, it becomes unnecessary to compute $Y_1^m, Y_2^m, \ldots, Y_s^m$ for any particular m as an exact solution to (2.7) as long as convergence of the overall scheme to an acceptable solution to (2.2) is assured.

Along with the scheme (2.8) we will also consider a slightly modified iteration method in which the "defect part" of (2.8) depends not only on $Y_j^{(m-1)} - y_{n-1}$ $(j = 1,2,\ldots,s)$ but also on $hf(Y_j^{m-1})$, $(j = 1,2,\ldots,s)$. Rather than use the full generality of the resulting scheme, we will consider only the extreme case in which the defect part of (2.8) depends *only* on derivative terms and thus takes the form

$$(2.9) \qquad Y_i^m = y_{n-1} + h \sum_{j=1}^{i} \bar{a}_{ij} f(Y_j^m) + h \sum_{j=1}^{s} w_{ij} f(Y_j^{m-1})$$

where, for consistency, the matrix W, with elements w_{ij} $(i,j = 1,2,\ldots,s)$ satisfies

$$(2.10) \qquad \bar{A} + W = A.$$

Finally, we will consider an iteration scheme in which the crude method takes the form of a Rosenbrock rather than a Runge-Kutta method. In other words we will consider a scheme in which only a single iteration of each stage of the crude method is performed. This approach, which seems to be the most promising of those we are discussing will be considered in detail in section 4 following a consideration of the schemes given by (2.8) and by (2.9) in section 3.

3. THE CHOICE OF $\bar{A}$

We will consider the effect of the choice of $\bar{A}$ on the rate of convergence of the overall scheme given by (2.8), taking into account the need for $\bar{A}$ to be of such a form as to simplify the computations.

For a linear test problem $y' = qy$ with $hq = z$, the vector $Y^m = [Y_1^m, Y_2^m, \ldots, Y_s^m]$ is given by

$$Y^m = (I - z\bar{A})^{-1} V \, Y^{m-1} + y_{n-1}(I - z\bar{A})^{-1}(I - V)e$$

where e is the vector with each of its s components equal to 1. We define the amplification factor of this iteration scheme for given z to be the spectral radius $\rho((I - z\bar{A})^{-1}V)$ and we wish to study the behaviour of this for z near zero and throughout the negative half of the complex plane.

In particular, we note that, if $\bar{A}$ is non-singular, then the amplification factor tends to zero as $z \to \infty$. Also, at $z = 0$, the amplification factor is $\rho(V)$ which equals the greatest distance between 1 and a point in $\sigma(\bar{A}A^{-1})$. Thus, the amplification factor is a measure of the quality of $\bar{A}$ as an approximation to A.

The first possibility we consider, is to require that V be upper triangular and that $\bar{A}$ (which of course is lower triangular) has constant diagonal elements. This can be achieved by taking $\bar{A}$ as kL^{-1} where k is a scalar constant and L is the unit lower triangular matrix in the LU factorization of A^{-1}. If the upper triangular factor U has diagonals $d_1, d_2, \ldots, d_s$, then $V = I - kU$ with eigenvalues $(1 - kd_i)$, $i=1,2,\ldots,s$. Choosing k as $2/(\max d_i + \min d_i)$ to minimize $\rho(V)$ gives a result

$$\rho(V) = (\max d_i - \min d_i)/(\max d_i + \min d_i).$$

For the well known 4th order 2 stage method

$$\begin{array}{c|cc} \frac{1}{2} - \frac{\sqrt{3}}{6} & \frac{1}{4} & \frac{1}{4} - \frac{\sqrt{3}}{6} \\ \frac{1}{2} + \frac{\sqrt{3}}{6} & \frac{1}{4} + \frac{\sqrt{3}}{6} & \frac{1}{4} \\ \hline & \frac{1}{2} & \frac{1}{2} \end{array}$$

we find that

$$A^{-1} = \begin{bmatrix} 3 & -3 + 2\sqrt{3} \\ -3 - 2\sqrt{3} & 3 \end{bmatrix} = \begin{bmatrix} 1 & 0 \\ -1 - \frac{2}{3}\sqrt{3} & 1 \end{bmatrix} \begin{bmatrix} 3 & -3 + 2\sqrt{3} \\ 0 & 4 \end{bmatrix}$$

so that

$$\bar{A} = \begin{bmatrix} \frac{2}{7} & 0 \\ \frac{2}{7} + \frac{4\sqrt{3}}{21} & \frac{2}{7} \end{bmatrix}, \quad V = \begin{bmatrix} \frac{1}{7} & \frac{6}{7} - \frac{4\sqrt{3}}{7} \\ 0 & -\frac{1}{7} \end{bmatrix}.$$

A numerical search shows that $\rho(V) = 1/7$ is slightly exceeded by $\rho((I - z\bar{A})^{-1}V)$ for some z values on the imaginary axis but the convergence rate is quite satisfactory throughout the negative half plane.

With $\bar{A}$ chosen in this way, the situation is not quite so acceptable for the 6th order 3 stage method for which it is found that the optimal value of k gives a value $\rho(V) = 29/79 \approx .37$.

On the other hand, a considerable improvement is achieved by renumbering the stages so that the method is written

$$\begin{array}{c|ccc}
\frac{1}{2}-\frac{\sqrt{15}}{10} & \frac{5}{36} & \frac{5}{36}-\frac{\sqrt{15}}{30} & \frac{2}{9}-\frac{\sqrt{15}}{15} \\[2ex]
\frac{1}{2}+\frac{\sqrt{15}}{10} & \frac{5}{36}+\frac{\sqrt{15}}{30} & \frac{5}{36} & \frac{2}{9}+\frac{\sqrt{15}}{15} \\[2ex]
\frac{1}{2} & \frac{5}{36}+\frac{\sqrt{15}}{24} & \frac{5}{16}-\frac{\sqrt{15}}{24} & \frac{2}{9} \\
\hline
\end{array}$$

to give

$$A^{-1} = \begin{bmatrix} 5 & 5-\frac{4}{3}\sqrt{15} & -4+\frac{4}{3}\sqrt{15} \\ 5+\frac{4}{3}\sqrt{15} & 5 & -4-\frac{4}{3}\sqrt{15} \\ -\frac{5}{2}-\frac{5}{6}\sqrt{15} & -\frac{5}{2}+\frac{5}{6}\sqrt{15} & 2 \end{bmatrix}$$

$$= \begin{bmatrix} 1 & 0 & 0 \\ 1+\frac{4}{15}\sqrt{15} & 1 & 0 \\ -\frac{1}{2}-\frac{1}{6}\sqrt{15} & -\frac{5}{8}+\frac{3}{16}\sqrt{15} & 1 \end{bmatrix} \begin{bmatrix} 5 & 5-\frac{4}{3}\sqrt{15} & -4+\frac{4}{3}\sqrt{15} \\ 0 & \frac{16}{3} & -\frac{17}{3}-\frac{8}{5}\sqrt{15} \\ 0 & 0 & \frac{9}{2} \end{bmatrix}.$$

If the optimal choice $k = 2/(\frac{16}{3}+\frac{9}{2}) = 12/59$ is made, it is now found that

$$\bar{A} = \begin{bmatrix} \frac{12}{59} & 0 & 0 \\ -\frac{12}{59}-\frac{16}{25}\sqrt{15} & 1 & 0 \\ \frac{15}{118}+\frac{9}{230}\sqrt{15} & \frac{15}{118}-\frac{9}{236}\sqrt{15} & 1 \end{bmatrix}, \quad V = \begin{bmatrix} \frac{-1}{59} & -\frac{60}{59}+\frac{16}{59}\sqrt{15} & \frac{48}{59}-\frac{16}{59}\sqrt{15} \\ 0 & \frac{-5}{59} & \frac{64}{59}+\frac{96}{295}\sqrt{15} \\ 0 & 0 & \frac{5}{59} \end{bmatrix}$$

giving a value $\rho(V) = \frac{5}{59} \approx .085$. In the search for more quickly convergent schemes at the expense of the computational convenience afforded by V being upper triangular, we next consider $\bar{A}$ chosen with equal diagonals and lower triangular form but so constrained as to ensure that $\rho(V) = \rho(I - \bar{A}A^{-1}) = 0$. This condition implies that $\det(\bar{A}) = \det(A)$ so that the common value of the diagonal elements of $\bar{A}$ is $\det(A)^{1/s}$. In the case $s = 2$ it is necessary and sufficient that $\bar{a}_{21}$

be chosen so that in addition to $\det(\bar{A}) = \det(A)$, it holds that $\mathrm{tr}(\bar{A}A^{-1}) = 2$. For the special case with order 4 we find that

$$\bar{A} = \begin{bmatrix} \frac{\sqrt{3}}{6} & 0 \\ \frac{\sqrt{3}}{3} & \frac{\sqrt{3}}{6} \end{bmatrix}, \quad V = \begin{bmatrix} 1 - \frac{\sqrt{3}}{2} & -1 + \frac{\sqrt{3}}{2} \\ 1 - \frac{\sqrt{3}}{2} & -1 + \frac{\sqrt{3}}{2} \end{bmatrix}$$

For $s = 3$, besides choosing the diagonals of $\bar{A}$ so that $\det(\bar{A}) = \det(A)$, the values of $\bar{a}_{21}$, $\bar{a}_{31}$, $\bar{a}_{32}$ must satisfy the requirements that $\mathrm{tr}(\bar{A}A^{-1}) = \mathrm{tr}((\bar{A}A^{-1})^2) = 3$. A computer evaluation shows that this can be achieved in the case of the order 6 method with

$$\bar{a}_{11} = \bar{a}_{22} = \bar{a}_{33} = 120^{-1/3} \approx .202700665191$$

$$\bar{a}_{21} \approx .355780210607$$

$$\bar{a}_{31} \approx -.9344818970137$$

$$\bar{a}_{32} = 0$$

We now return to the scheme given by (2.9) where W is given by (2.10). As before we consider the linear test problem $y' = qy$ with $hq = z$. The iteration scheme is

$$Y^m = ((I - z\bar{A})^{-1}zW)Y^{m-1} + y_{n-1}(I - z\bar{A})^{-1}e$$

with amplification factor $\rho((I - z\bar{A})^{-1}zW)$ which vanishes for $z = 0$. On the other hand, as $z \to \infty$, we have the limiting value $\rho(\bar{A}^{-1}W) = \rho(\bar{A}^{-1}(A - \bar{A})) = \rho(\bar{A}^{-1}A - I)$. If we wish $\rho(\bar{A}^{-1}A - I)$ to vanish exactly, it is interesting that the same restriction is placed on $\bar{A}$ as in the previous scheme since $\sigma(\bar{A}^{-1}A) = \{1\}$ is equivalent to $\sigma(A^{-1}\bar{A}) = \{1\}$ and since $\sigma(\bar{A}A^{-1}) = \sigma(A^{-1}\bar{A})$.

For the method of order 4 with $s = 2$ we achieve this optimal convergence situation with

$$\bar{A} = \begin{bmatrix} \frac{\sqrt{3}}{6} & 0 \\ \frac{\sqrt{3}}{3} & \frac{\sqrt{3}}{6} \end{bmatrix}, \quad W = \begin{bmatrix} \frac{1}{4} - \frac{\sqrt{3}}{6} & \frac{1}{4} - \frac{\sqrt{3}}{6} \\ \frac{1}{4} - \frac{\sqrt{3}}{6} & \frac{1}{4} - \frac{\sqrt{3}}{6} \end{bmatrix}$$

For other values of z we can compute $\rho((I - z\bar{A})^{-1}zW)$ very simply as $|tr((I - z\bar{A})^{-1}zW)|$ since the trace of a singular 2×2 matrix gives the value of its only non-zero eigenvalue. This gives a result $|z(\frac{1}{2}-\frac{\sqrt{3}}{3})/(1 - z\frac{\sqrt{3}}{6})^2|$ which is precisely the same as $|tr((I - z\bar{A})^{-1}V)|$ resulting from the previous form of the iteration scheme for this method. The maximal value of the amplification factor in the negative half plane turns out to be $1 - \sqrt{3}/2 \approx .134$ when $z = \pm 2\sqrt{3}\ i$.

4. SCHEMES WITH IMPROVED STORAGE EFFICIENCY

Rather than building an iteration scheme on a crude method of semi-explicit type, we now consider a modification in which only a single iteration in each stage is carried out. With this change, all of the previously considered schemes have exactly the same convergence behaviour under the linear test problem but the underlying crude method assumes Rosenbrock form.

To make this Rosenbrock nature of the iteration scheme more overt, we consider a rewriting of it. As usual, let $Y_1^{m-1}, Y_2^{m-1}, \ldots, Y_s^{m-1}$ be the results computed in iteration number m-1 and define $R_1^{m-1}, R_2^{m-1}, \ldots, R_s^{m-1}$ as the residuals for equation (2.2), scaled by $(I - \lambda hJ)^{-1}$ where λ is a real number and J is a recently computed value of the Jacobian. Thus,

$$R_i^{m-1} = (I - \lambda hJ)^{-1}(Y_i^{m-1} - y_{n-1} - h \sum_{j=1}^{s} a_{ij} f(Y_j^{m-1})),$$

and it is proposed to define $Y_1^m, Y_2^m, \ldots, Y_s^m$ by

$$Y_i^m = Y_i^{m-1} - \sum_{j=1}^{s} p_{ij} R_j^{m-1}$$

for $p_{11}, \ldots, p_{ss}$ elements of some non-single matrix P. It is clear that the limit to the sequence generated by this iteration, if this limit exists, satisfies (2.2). As for the convergence rate, we study this as usual using a linear model $y' = qy$ with $z = hq$, to obtain the iteration matrix

$$I - \frac{1}{1 - \lambda z} (P - zPA)$$

which has value $I - P$ when $z = 0$ and $I - \frac{1}{\lambda} PA$ when $z = \infty$. To obtain better than linear convergence at these two extremes, we can choose $\lambda = \det(A)^{1/s}$ and P as a unit lower triangular matrix such that

$\sigma(\frac{1}{\lambda} PA) = \{1\}$. An alternative is to choose $P = QA^{-1}$ with Q a lower triangular matrix with constant diagonals equal to $\lambda = \det(A)^{1/s}$ such that $\sigma(QA^{-1}) = \{1\}$.

Both of these choices require the storage of at least 2s vectors throughout an integration step.

To see how this storage requirement can be reduced, we write the second alternative in the form

$$S_i^{m-1} = (I - \lambda hJ)^{-1}\left(\sum_{j=1}^{s} a_{ij}^{(-1)}(Y_j^{m-1} - y_{n-1}) - hf(Y_i^{m-1})\right)$$

$$Y_i^m = Y_i^{m-1} - \sum_{j=1}^{s} q_{ij} S_j^{m-1} ,$$

where $a_{ij}^{(-1)}$ is an element of A^{-1}, and replace the term

$$\sum_{j=1}^{s} a_{ij}^{(-1)}(Y_j^{m-1} - y_{m-1})$$

by

$$\sum_{j<i} a_{ij}^{(-1)}(Y_j^m - y_{n-1}) + \sum_{j\geq i} a_{ij}^{(-1)}(Y_j^{m-1} - y_{n-1})$$

so that the newly computed values of the iterates can overwrite the old ones as they are evaluated. Let $L + U = A^{-1}$ where L is strictly lower triangular and U is upper triangular. In this case the iteration matrix for the standard linear problem is

$$I - ((1 - \lambda z)Q^{-1} + L)^{-1}(A^{-1} - zI)$$

with value $I - (Q^{-1} + L)^{-1}A^{-1}$ when $z = 0$ and $I - \frac{1}{\lambda}Q$ when $z = \infty$. To obtain faster than linear convergence at each of these points, we may choose Q as a lower triangular matrix with diagonal elements equal to $\lambda = \det(A)^{1/s}$ and such that $\sigma((Q^{-1} + L)A) = \{1\}$.

As it happens, it is possible to choose Q satisfying this requirement for the methods of order 2s for s = 1,2,...,5 with only the diagonals and first column of Q non-zero. For these methods, the first column of Q is as follows

$s = 2$: $q_{11} = \sqrt{3}/6 = .288675134595$, $q_{21} = \sqrt{3}/6 - 1/4 = .0386751345948$;

$s = 3$: $q_{11} = .202740066191$, $q_{21} = .02734550512951$,

$q_{31} = .007382171935621$;

$s = 4$: $q_{11} = .156196996846$, $q_{21} = -.0552894874522$,

$q_{31} = -.314665006593$, $q_{41} = -.4725442471881$;

$s = 5$: $q_{11} = .127023373512$, $q_{21} = -.0796654991277$,

$q_{31} = -.1095675191942$, $q_{41} = 1.23636819435$,

$q_{51} = 3.5976266823$.

5. MISCELLANEOUS COMMENTS

In this paper, all the examples have been based on the methods of optimal order. The main reason for making this choice is that other approaches work poorly for these methods. For example, transformation to nearly diagonal form is of little use because the eigenvalues of the A matrix are all distinct and at most one is real.

However, these methods are potentially very useful because they have an h^2 error expansion and are thus eminently suitable for the application of extrapolation techniques.

Preliminary investigations indicate that it may be possible to combine into one itegration step a step of a method and two further half steps of the same method and to construct iteration schemes of the style we have been discussing with a single $I - h\lambda J$ factorization needed for the composite method.

A technique introduced into the STRIDE algorithm for extending the range of step sizes for which a stale $I - h\lambda J$ factorization may be usefully retained is applicable to some of the methods in this paper. In this technique, if h_0 was the step size when $I - h\lambda J$ was evaluted and h_1 is a current value, then a relaxation factor $2/(1 + h_1/h_0)$ is applied to the modified Newton iterations. We note that for linear multistep methods, as well as for the singly implicit methods used in STRIDE, use of this factor does the least harm to the convergence rate for approximately linear solution components varying from zero to infinite stiffness. Use of this relaxation factor has a similar effect to the methods in §3 which have super-linear convergence at zero and infinity.

Finally, limited numerical testing of the methods of §4, shows that they are capable of achieving comparable speeds of convergence to that yielded by a straightforward modified Newton method. For low values of s and for low precision requirements their behaviour is in fact almost identical but as s and the precision increase they seem to deteriorate. There seems to be a lot more to learn about these iteration schemes: different permutations of the stages can drastically change the speed of convergence.

ACKNOWLEDGEMENTS

My interest in this topic came largely from my collaboration with Fred Chipman and Kevin Burrage in the development of STRIDE. I am most grateful for this association. The writing of this paper was completed at the Mathematics Institute of Linköping University. I wish to express my appreciation to the Numerical Analysis group and others in this institute for the supportive and stimulating scientific atmosphere I enjoyed there.

REFERENCES

1. Alexander, R. "Diagonally implicit Runge-Kutta methods for stiff o.d.e.'s", SIAM J. Numer. Anal., 14, 1006-1021 (1977).

2. Bickart, T.A. "An efficient solution process for implicit Runge-Kutta methods", SIAM J. Numer. Anal., 14, 1022-1027 (1977).

3. Burrage, K. "A special family of Runge-Kutta methods for solving stiff differential equations", BIT, 18, 22-41 (1978).

4. Burrage, K., Butcher, J.C., and Chipman, F.H. "An implementation of singly-implicit Runge-Kutta methods", Comp. Maths. Report, Univ. of Auckland (in preparation).

5. Butcher, J.C. "On the implementation of implicit Runge-Kutta methods", BIT, 16, 237-240 (1976).

6. Butcher, J.C. "A transformed implicit Runge-Kutta method", Comp. Maths. Report No. 13, Univ. of Auckland (1977).

7. Butcher, J.C., Burrage, K., and Chipman, F.H. "STRIDE: stable Runge-Kutta integrator for differential equations", Comp. Maths. Report, Univ. of Auckland (in preparation).

8. Chipman, F.H. "The implementation of Runge-Kutta implicit processes", BIT, 13, 391-393 (1973).

9. Frank, R., and Überhuber, C.W. "Iterated defect correction for the efficient solution of stiff systems of ordinary differential equations", BIT, 17, 146-159 (1977).

10. Frank, R., and Überhuber, C.W. "Collocation and iterated defect correction", Lecture notes in Mathematics 631, 19-34, Springer-Verlag (1978).

11. Nørsett, S.P. "Semi-explicit Runge-Kutta methods", Math. and Comp. Report No. 6/74, Univ. of Trondheim (1974).

BEST APPROXIMATION IN TENSOR PRODUCT SPACES

E. W. Cheney‡

1. Introduction

This branch of approximation theory has come into existence in order to treat a number of practical problems involving multivariate functions. A significant feature of these problems is the presence of infinite-dimensional linear subspaces which serve as the sets of candidates for approximation.

A simple example of this occurs when it is desired to approximate a continuous function of two variables $f(x,y)$ by a sum of continuous univariate functions:

$$f(x,y) \approx r(x) + s(y) \ .$$

If f is defined on the square $0 \leq x \leq 1$, $0 \leq y \leq 1$, then r and s are functions which may range over $C[0,1]$. The totality of bivariate functions having the form $r(x) + s(y)$ is an infinite-dimensional linear subspace in $C([0,1] \times [0,1])$. This linear subspace is closed, and even proximinal. That means that each function in the underlying space possesses at least one best approximation in the subspace. Moreover, a characterization theorem is available for best approximations, and an efficient algorithm exists for computing them.

The ideal situation which exists for the type of approximation just described does not persist, however, for other similar problems. In the following sections we will outline the current state of affairs and point out important areas in which it is hoped others will be induced to take up investigations.

2. The Tensor-product Formalism

The approximation problems considered here are most easily interpreted with the ideas of tensor-products.

If E and F are two linear spaces, their tensor product is a new linear space denoted by $E \otimes F$ which consists of all formal expressions $\sum_{i=1}^{n} f_i \otimes g_i$, with $n \in \mathbb{N}$, $f_i \in E$, and $g_i \in F$. The set $E \otimes F$ is made into a linear space by defining addition,

‡Work supported in part by the U. S. Army Scientific Research Office at The University of Texas, Austin, Texas, U.S.A.

scalar multiplication, and equality appropriately. The most elementary discussion of these matters is in Schatten's monograph [1]. Consult also appropriate sections of [2], [3].

In the general theory, the tensor product construction is important because of the various interpretations to which the objects $\sum f_i \otimes g_i$ are subject. They can be interpreted as bilinear forms or as linear transformations, for example. The theory is also enriched by the fact that many useful norms can be introduced into the linear space $E \otimes F$. In our applications, f_i and g_i will be functions, and $f_i \otimes g_i$ will denote an ordinary product.

Suppose that E and F are linear spaces with norms. A norm α on $E \otimes F$ is called a <u>cross-norm</u> if $\alpha(f \otimes g) = \|f\| \cdot \|g\|$. Among the cross-norms, the following one is especially important:

$$\lambda(\sum f_i \otimes g_i) = \sup_{\substack{\phi \in E^* \\ \|\phi\|=1}} \sup_{\substack{\psi \in F^* \\ \|\psi\|=1}} \sum \phi(f_i)\psi(g_i) .$$

In this equation, E^* and F^* are the Banach-space conjugates of E and F.

The linear space $E \otimes F$, with norm α, is of course a metric space. The <u>completion</u> of that metric space is denoted by $E \otimes_\alpha F$.

If E and F are spaces of real-valued functions on domains X and Y, respectively, then $\sum_{i=1}^n f_i \otimes g_i$ has a natural interpretation as a function on $X \times Y$, namely,

$$(x,y) \longmapsto \sum_{i=1}^n f_i(x) g_i(y) .$$

Under this interpretation $C(X) \otimes_\lambda C(Y)$ is $C(X \times Y)$. We understand by $C(X)$ the space of continuous real functions on a (compact Hausdorff) space X normed by $\|f\| = \max |f(x)|$. In a similar way one can identify $L_1(X \times Y)$ with $L_1(X) \otimes_\Upsilon L_1(Y)$, when X and Y are measure spaces of finite measure, and Υ is another cross-norm. The advantage of making these interpretations is that the <u>general</u> theory yields at once such specialized theorems as the following one.

<u>THEOREM</u>. Let G and H be finite-dimensional subspaces of $C(X)$ and $C(Y)$, respectively. Then

$$G \otimes C(Y) + H \otimes C(X)$$

is a closed subspace of $C(X \times Y)$.

Such a theorem guarantees, for example, that the set of bivariate functions

$$\sum_{i=0}^n s_i(y) T_i(x) + \sum_{i=0}^m r_i(x) T_i(y)$$

is closed. Here n and m are fixed, T_i denotes a Tchebycheff polynomial, and the functions s_i and r_i range freely over $C[-1,1]$.

I do not know whether the subspace in the theorem is proximinal except in the case that G and H are 1-dimensional subspaces generated by functions without zeros. The proximinality of the latter has been proved by John Respess and myself (unpublished).

3. Applications to Integral Equations

We turn now to some practical problems of numerical analysis in which tensor-product approximations are used or can be used to advantage.

A familiar example occurs in the theory of integral equations (and therefore also in two-point boundary-value problems). Consider the integral equation

$$(1) \qquad x(t) = u(t) + \int_0^1 K(s,t)x(s)\,ds$$

in which x is an unknown function, while u and K are known. Suppose that the kernel K is replaced by an approximation k having the form

$$(2) \qquad k(s,t) = \sum_{i=1}^{n} g_i(s)h_i(t).$$

The substitution of k for K in Eq. (1) leads to

$$(3) \qquad x(t) = u(t) + \sum_{j=1}^{n} h_j(t) \int_0^1 g_j(s)x(s)\,ds$$

which we write in the abbreviated form

$$(4) \qquad x = u + \sum_{j=1}^{n} \langle x,g_j\rangle h_j .$$

Clearly, this equation gives x explicitly once the inner-products $\langle x,g_j\rangle$ have been determined. These in turn can be obtained by solving the n equations which result from taking the inner-product of both sides of (4) with g_i. These equations are

$$(5) \qquad \langle x,g_i\rangle = \langle u,g_i\rangle + \sum_{j=1}^{n} \langle x,g_j\rangle\langle h_j,g_i\rangle .$$

If system (5) is non-singular, the unknowns $\langle x,g_i\rangle$ can be obtained from it and substituted in Eq. (4).

This general procedure can have many concrete realizations depending upon the method employed to obtain the approximation k in Eq. (2). An interesting proposal made by N. M. Wang and reported by Gordon [4] is to select k so that we have interpolation of K along a network of horizontal and vertical <u>lines</u>. This is conveniently done by means of <u>blending</u> operators. For example, suppose that a partition

$$0 = \xi_0 < \xi_1 < \ldots < \xi_m = 1$$

is prescribed. Let the corresponding Lagrange operator be

$$(Lx)(s) = \sum_{i=0}^{m} x(\xi_i)\ell_i(s) \qquad \ell_i(s) = \prod_{\substack{j=0\\ j\neq i}}^{m} \frac{s-\xi_j}{\epsilon_i-\xi_j} .$$

This gives rise to two bivariate operators

(6) $$(Pf)(s,t) = \sum_{i=0}^{m} f(\xi_i, t)\ell_i(s)$$

(7) $$(Qf)(s,t) = \sum_{i=0}^{m} f(s,\xi_i)\ell_i(t) .$$

Then for any f defined on the square $0 \leq s,\ t \leq 1$, Pf interpolates to f on the lines $s = \xi_i$, while Qf interpolates to f on the lines $t = \xi_i$. As is well known, the Boolean sum

(8) $$(P \oplus Q)f = Pf + Qf - PQf$$

will interpolate to f on the network of all the given horizontal and vertical lines. Moreover, the function in Eq. (8) has the form of k in Eq. (2), as can be seen from Eq. (6) and (7).

The method just outlined should be contrasted with the more usual approach of fitting k to K on a network of points. In that case the form of k would be $\sum p_i(t)q_i(s)$ with p_i and q_i polynomials or splines. Of the many references possible to this procedure we cite two, [14] and [15].

The fitting of the approximate kernel k to the given kernel K implies that the solution in Eq. (4) is only approximate. Now Eq. (4) can be written in operator form as $x = u + \bar{K}x$, where $\bar{K}$ denotes the integral operator. Thus x should be a fixed point of the mapping $x \mapsto u + \bar{K}x$. Similarly, the approximate solution should be a fixed point of the mapping $x \mapsto u + \bar{k}x$. These mappings should be close to each other, and if the operators are interpreted as acting on the space C[0,1] with supremum norm, one is led to the problem of choosing the kernel k to minimize

(9) $$|||\bar{K} - \bar{k}||| = \sup_t \int_0^1 |K(s,t) - k(s,t)|ds .$$

This rather special approximation problem has been studied in [16] and [17]. Notice that Eq. (9) involves both a supremum norm and an L_1-norm.

4. An Application to Scaling of Matrices

In some problems of numerical linear algebra it is desirable to replace a matrix (a_{ij}) by a "scaled-copy." This is a matrix obtained from (a_{ij}) by dividing row i by r_i and column j by c_j $(1 \leq i \leq n,\ 1 \leq j \leq n)$. The new matrix (b_{ij}) is given by

(1) $$b_{ij} = a_{ij}/r_ic_j .$$

Scaling is carried out in order to improve the conditioning of problems involving (a_{ij}). A convenient criterion might be to achieve a minimum in the expression

(2) $$\max_{ij} |b_{ij}| / \min_{ij} |b_{ij}|.$$

Using the principle of homogeneity, we see that what is desired is to achieve the bounds

(3) $$1 \leq |b_{ij}| \leq k$$

with a minimum value for k. If capital letters are used to denote the logarithms of magnitudes of the variables, we obtain

(4) $$0 \leq B_{ij} = A_{ij} - R_i - C_j \leq K .$$

For simplicity one should assume here that all entries a_{ij} are nonzero. The objective is to select R_i and C_j so that K is a minimum in (4). The solution is obtained by minimizing

(5) $$\epsilon = \|A - R - C\|_\infty = \max_{ij} |A_{ij} - R_i - C_j| .$$

Then we have

(6) $$2\epsilon \geq A_{ij} - R_i - C_j + \epsilon \geq 0$$

so that $R_i - \epsilon$ and C_j solve the problem (4). The approximation problem (5) can be solved by the Diliberto-Straus algorithm [5], although this fact was not noticed in [6]. R. Bank has exploited these ideas to accelerate the convergence of iterative schemes for elliptic partial differential equations. See [7, 8, 9]. Also, as was pointed out to me by Dr. Eugene Wachspress, a similar additive pre-conditioning of matrices is the basis for the "two-dimensional correction method" of Settari and Aziz [10]. Two recent papers by von Golitschek are pertinent here [11], [12], as well as the report by Rothblum and Schneider [13].

5. Status of the Best Approximation Problem

We consider L_2-spaces first, since the results there are elementary. Suppose that we wish to approximate a given function f(x,y) in the L_2-norm by a function from the subspace

(1) $$U = G \otimes L_2(Y) + H \otimes L_2(X)$$

Such a function will have the form

(2) $$u(x,y) = \sum_{i=1}^{n} r_i(y) g_i(x) + \sum_{i=1}^{m} s_i(x) h_i(y)$$

where $\{g_1,\ldots,g_n\}$ is a fixed basis for G and $\{h_1,\ldots,h_m\}$ a fixed basis for H. The functions r_i and s_i, however, range freely over $L_2(Y)$ and $L_2(X)$, respectively.

The solution to this problem is immediate. Assume that the bases chosen for G and H are orthonormal. Define orthogonal projection operators P, Q, and B by the formulas

(3) $$(Pf)(x,y) = \sum_{i=1}^{n} g_i(x) \int_X f(s,y)g_i(s)\,ds$$

(4) $$(Qf)(x,y) = \sum_{i=1}^{m} h_i(y) \int_Y f(x,t)h_i(t)\,dt$$

(5) $$B = P + Q - PQ\ .$$

Then $u = Bf$ is the function sought. It can be seen that P and Q are the orthogonal projections onto $G \otimes L_2(Y)$ and $H \otimes L_2(X)$, respectively. The operator B is the Boolean sum of P and Q and is the orthogonal projection onto U. Thus B produces a best approximation at once, and the process is linear.

In spaces other than Hilbert space, the best approximation operators P and Q are nonlinear. They are idempotent but can be discontinuous. The Boolean sum now must be defined as $B = P + Q(I-P)$ since linearity is not present. The operator B will generally not produce best approximations at once. An iterative scheme that works in some cases consists in computing the sequence

(6) $$u_n = f - (I-B)^n f\ .$$

An easy inductive proof shows that $u_n \in U$ and that $\|f-u_n\| \leq \|f-u_{n-1}\|$. In spaces having some smoothness, like L_p $(1 < p < \infty)$, Atlestam and Sullivan [18], [19] have established the convergence of the sequence $\{u_n\}$ to a best approximation of f. In [25], Golomb proves convergence of $\|f-u_n\|$ to dist(f,U) under rather restrictive conditions, which apparently are rarely fulfilled. His theorem does apply to the case originally considered by Diliberto and Straus, namely

(7) $$U = \pi_0(X) \otimes C(Y) + \pi_0(Y) \otimes C(X)\ .$$

Here π_0 denotes the space of constant functions. In this case not only does $\|f-u_n\|$ converge to dist(f,U) but the functions u_n themselves converge uniformly, as first proved by Aumann [20] and later by others unaware of Aumann's work [21] and [22].

It is natural to attempt the same iteration for a slightly more general space, for example

(8) $$U = \pi_1(X) \otimes C(Y) + \pi_0(Y) \otimes C(X)$$

but here the algorithm may fail, as discovered by Dyn [23]. A satisfactory algorithm is therefore not available for the best approximation problem in this case.

The same situation prevails in the L_1-space. Here we use as approximating subspace

(9) $$U = G \otimes L_1(Y) + H \otimes L_1(X)$$

considered as a subspace in $L_1(X \times T)$. Unpublished work of W. A. Light and myself establishes the existence of best approximations to each essentially bounded function in $L_1(X \times Y)$ provided the finite-dimensional subspaces G and H consist of essentially bounded functions. Here X and Y are measure spaces of finite measure. Some earlier results in this direction appear in [24].

The state of affairs can be summarized by saying that in the simplest case (Eq. 7) the Diliberto-Straus Algorithm works very well and converges. In all other cases, one probably should make do with a simple blending projection such as the one described by Eq. (3, 4, 5) above or the interpolating projection described by Eq. (6, 7, 9) in Section 3.

References

1. Schatten, Robert, "A Theory of Cross-spaces," Annals of Mathematics Studies, No. 26, Princeton University Press, Princeton, 1950.
2. Robertson, A.P., and Wendy Robertson, "Topological Vector Spaces," Cambridge University Press, Cambridge, 1964.
3. Schaeffer, H. H., "Topological Vector Spaces," MacMillan Company, New York, 1966.
4. Gordon, W. J., "Blending-function methods of bivariate and multivariate interpolation and approximation," General Motors Research Laboratories, Report 834, October 1968.
5. Diliberto, S. P., and E. G. Straus, "On the approximation of a function of several variables by the sum of functions of fewer variables," Pacific J. Math. 1 (1951), 195-210.
6. Fulkerson, D. R., and P. Wolfe, "An algorithm for scaling matrices," SIAM Review 4 (1962), 142-146.
7. Bank, R., "An automatic scaling procedure for a d'Yakanov-Gunn iteration scheme," CNA Report 142, University of Texas at Austin, August 1978. To appear, Linear Algebra and Its Applications.
8. Bank, R., and D. J. Rose, "Marching algorithms for elliptic boundary value problems I: The constant coefficient case," SIAM J. on Numerical Analysis 14 (1977), 792-829.
9. Bank, R., "Marching algorithms for elliptic boundary value problems II: The variable coefficient case," SIAM J. on Numerical Analysis 14 (1977), 950-970.
10. Settari, A., and K. Aziz, "A generalization of the additive correction methods for the iterative solution of matrix equations," SIAM J. on Numerical Analysis 10 (1973), 506-521.
11. v. Golitschek, M., "An algorithm for scaling matrices and computing the minimum cycle mean in a digraph," preprint, August 1979, Institute for Applied Mathematics, University of Würzburg, Germany.
12. v. Golitschek, M., "Approximation of functions of two variables by the sum of two functions of one variable," preprint, August 1979, Institute for Applied Mathematics, University of Würzburg, Germany.
13. Rothblum, U. G., and H. Schneider, "Characterizations of optimal scalings of matrices," Report RS 2678 (1978), Yale University, New Haven.

14. Hammerlin, G., and L. L. Schumaker, "Procedures for kernel approximation and solution of Fredholm integral equations of the second kind," CNA Report 1 8, University of Texas at Austin, November 1977.

15. Atkinson, K., "A Survey of Numerical Methods for the Solution of Fredholm Integral Equations of the Second Kind," SIAM, Philadelphia, 1976.

16. Phillips, G. M., J. H. McCabe, E. W. Cheney, "A mixed-norm bivariate approximation problem with applications to Lewanowicz operators," pp. 315-323 in "Multivariate Approximation," D.C. Handscomb, ed., Academic Press, New York, 1978. Also CNA Report 127, University of Texas at Austin, October 1977.

17. Krabs, W., "Optimierung und Approximation," B. G. Teubner, Stuttgart, 1975 (particularly Sec. 5.3).

18. Atlestam, B., and F. E. Sullivan, "Iteration with best approximation operators," Rev. Roumaine Math. Puras Appl. 21 (1976), 125-131. MR 53#6188.

19. Sullivan, F. E., "A generalization of best approximation operators," Ann. Mat. Pura Appl. 107 (1975), 245-261. MR 53#3564.

20. Aumann, G., "Uber Approximative Nomographie II," Bayer. Akad. Wiss. Math. Nat. Kl. S.B. 1959, 103-109. MR 22#6968.

21. Light, W. A., and E. W. Cheney, "On the approximation of a bivariate function by the sum of univariate functions," to appear, J. Approximation Theory. Also CNA Report 140, University of Texas at Austin, August 1978.

22. Kelley, C. T., "A note on the approximation of functions of several variables by sums of functions of one variable," Report 1873, Math. Research Center, Madison, Wisconsin, August 1978.

23. Dyn, R., "A straightforward generalization of Diliberto and Straus' algorithm does not work," to appear, J. Approximation Theory. Report dated December 1978, Math. Research Center, Madison, Wisconsin.

24. Light, W. A., J. H. McCabe, G. M. Phillips, and E. W. Cheney, "The approximation of bivariate functions by sums of univariate ones using the L-1 metric," Report CNA 147, University of Texas at Austin, March 1979.

25. Golomb, M., "Approximation by functions of fewer variables," pp. 275-327 in "On Numerical Approximation," R. E. Langer, ed., University of Wisconsin Press, 1959.

MONOTONICITY AND FREE BOUNDARY VALUE PROBLEMS

L. Collatz

Summary: Free boundary value problems were treated extensively in the last years, mostly by the help of discretization methods; here monotonicity properties are used, which give (in not too complicated cases) inclusion theorems for the free boundaries. This is illustrated in some examples, even in free boundary value problems in more dimensions.

Introduction:

A classical physical phenomenon is described by the Latin word "actio et reactio", which means, that often enlarging the forces causes enlarging the influence. In many cases this is mathematically equivalent to the principle of monotonicity, and gives the possibility of inclusion theorems, as it will be illustrated by the following examples. Very often, the principle of monotonicity is physically evident, but mathematically difficult to prove. If there exists a mathematical proof, one gets lower and upper bounds for the free boundary which one can guarantee. If none exists one can also include the free boundary but without guarantee. Therefore it is very important to enlarge the field of applicability of monotonicity.

1. A simple model in one dimension.

An ideal rope in the x-y-plane is suspended at the point $x=0$, $y=y_0$, Fig. 1, and is lying partially on a horizontal table $y=0$. The graph $y(x)$ of the rope satisfies along $0<x<a$ the classical differential equation of a rope under the influence of the gravity working in opposite direction

to the y-axis

(1.1) $y'' = \gamma\sqrt{1+y'^2}$ $(0<x<a)$

where γ is a constant; furthermore we have

(1.2) $y(x) \equiv 0$ for $a\leq x\leq b$,

where b is determined by the given length ℓ of the rope. We have for the differential equation of second order (1.1) three boundary conditions:

(1.3) $y(0) = y_0$, $y(a) = 0$, $y'(a) = 0$;

but a is not known, a is the "free boundary". We can transform the problem into a problem with a fixed interval $(0,c)$, if we choose c sufficiently large, for instance $c > \ell$. Then we have the problem:

(1.4) $y'' = \gamma\sqrt{1+y'^2}$ for $0<x<a$; $y=0$ for $a<x<c$,

with the condition (1.3).

We observe, that the function $y''(x)$ is discontinuous at $x=a$. This discontinuity in the higher derivatives is typical for the free boundaries and causes certain difficulties. One can study the phenomenon a little easier by taking the linearized equation $y''=\gamma$ instead of (1.1).

Many free boundary value problems have been studied in the last decade; only a few examples may be mentioned: The classical (one phase and two phases) Stefan Problem Fig. 2 (compare Rubinstein [71] Baiocchi [72] Crank a.o. [75] Friedmann a.o. [75] Hoffmann [78], Miller a.o. [78], Wilson a.o.) contact problem (a cylinder lies on a beam and with respect to elastic deformations it has contact with the beam along a finite domain, Fig.3), elasto-plastic problem, Fig.4, concentration problems and many others.

2. Different formulations of free boundary value problems.

The following type of a free boundary value problem occurs in different applications: We have a domain G in the n-dimensional $x_1,\dots,x_n$-space with a fixed boundary Γ_1 and a free boundary Γ and the differential equation of second order for an unknown function $u(x)=u(x_1,\dots,x_n)$

(2.1) $$Lu = f(x) \text{ in } G$$

and the boundary conditions

(2.2) $$u(x)= k(x) \text{ on } \Gamma_1, \quad u = \frac{\partial u}{\partial n} = 0 \text{ on } \Gamma;$$

f and k are given continuous functions and $k(x) \geq 0, k \not\equiv 0$; n is the outer normal. We suppose that the classical maximum principle holds for the differential equation. Therefore we have $u(x) > 0$ in G. We define $u = 0$ outside of G. This is the first formulation I.

Now we assume that a priori we know a domain B which is so large, that is contains G; Fig.5; then we can formulate the problem in the form of a boundary value problem with fixed domain B:

(2.3) $$Lu \geq f \text{ in } B, \; u \geq 0 \text{ in } B, \; u = k(x) \text{ on } \Gamma_1.$$

With respect of the discontinuity in the higher derivatives we take the function u of the class $C^1(B)$ and a.e. in $C^2(B)$, (formulation II).

The formulation III is

(2.4) $$u(Lu-f) = 0 \text{ in } B, \text{ with the restrictions}$$

(2.5) $$Lu \geq f \text{ in } B, \quad u \geq 0 \text{ in } B, \; u = k(x) \text{ on } \Gamma_1, \; u \in C^1(B).$$

The equation $u(Lu-f) = 0$ in B is satisfied, if either $u>0$ and then $Lu=f$ (that is true in G) or $u=0$, this can happen only outside of G, because $u>0$ in G.

The formulation IV: uses a variational principle:

(2.6) $$I[u] = \int_B u(Lu-f)dx = \text{Min under the restrictions (2.5)}$$

One has $I[u] \geq 0$ for all admissable functions and $I[u] = 0$ for the solution u of (2.4) (2.5). In the special case $Lu = -\Delta u$ (=- Laplace-Operator), $f=0$ one has

(2.7) $$I[u] = \int_B uLu\,dx = \int_B (\text{grad } u)^2\,dx + \int_{\partial B} \dots$$

where the first integral on the right side is the well known Dirichlet-integral and the second integral is a boundary integral.

The variational principle allows the application of the classical methods of Ritz-Galerkin a.o.

3. Comparison-Theorem

(Compare Meyn-Werner [79] ; I thank Prof. Bather, Prof. Baiocchi and Prof. Bodo Werner for suggestions and discussions about this Nr.3).

Let us compare the following two free boundary value problems for two functions u and u^*:

(3.1) $$Lu=f \leq 0 \text{ in } G,\; u = k(x) \geq 0 \text{ on } \Gamma_1,\; u > 0 \text{ in } G,$$
$$u = \frac{\partial u}{\partial n} = 0 \text{ on } \Gamma,$$

(3.2) $$Lu^* = f^* \leq 0 \text{ in } G^*,\; u^*>0 \text{ in } G^*,\; u=k^* \geq 0 \text{ on } \Gamma_1,$$
$$u^* = \frac{\partial u^*}{\partial n} = 0 \text{ on } \Gamma^*.$$

Both problems may satisfy the conditions of the class of problems in Nr. 2 and may have uniquely determined solutions u, resp. u^*. Furthermore we suppose:

1) L is a homogeneous linear operator, mapping the zero-function into itself:

$$L0 = 0,$$

2) L admits a monotonicity principle for every open bounded domain H with boundary ∂H:

(3.3) From $Lu \geq 0$ in H, $v(x) \geq 0$ on ∂H follows $v(x) \geq 0$ in H.

We assume, that there exists a domain B which contains G and G^*, Fig.6, and we will use the equivalence between the different formulations of free boundary value problems in Nr.2. We suppose

(3.4) $f^*(x) \geq f(x)$, $k^*(x) \geq k(x)$ and we will show $u^* \geq u$ in B.

We consider the difference

$$z(x) = u(x) - u^*(x)$$

We have in the domain $\hat{B} = B - (G \cup G^*)$, Fig.6,

(3.5) $u = u^* = 0$, therefore $z = 0$ in $\hat{B}$.

$\tilde{G}$ may be the open **domain** $\tilde{G} = \{x \in B,\ z > 0\}$.
With respect to $k^* \geq k$ is

(3.6) $z \leq 0$ on Γ_1 or

(3.7) $z \leq 0$ in $B - \tilde{G}$.

On the boundary holds

$$z = 0 \text{ on } \partial\tilde{G} \cap B.$$

Therefore (3.8) $z \leq 0$ on the whole boundary $\partial\tilde{G}$.

In $\tilde{G}$ we have $u^* \geq 0$, $u = z + u^* > 0$.

The formulation III in Nr. 2 gives (with respect to $u > 0$)

(3.9) $\quad Lu = f$ in $\tilde{G}$.

In G we have either $u^* = 0$ and then $Lu^* = 0$ or $u^* > 0$ and then $Lu^* = f^* \leq f$; in both cases

(3.1o) $\quad Lz = Lu - Lu^* = f \leq 0$ in G.

With respect to the monotonicity principle (3.3) the consequence is

(3.11) $\quad Z \leq 0$ in $\tilde{G}$. This contradicts $z > 0$ in $\tilde{G}$,

therefore $\tilde{G}$ is empty, $u^* \geq u$ in the whole domain B or the domain G is contained in the domain G^*.
This gives the possibility: inclusions of free boundaries: The problem (3.1) is to be solved with given Γ_1, f, k. If one can find a function u^* that solves (3.2) with certain f^*, k^* satisfying (3.4), then the corresponding domain G^* contain G, and we have got an upper bound for Γ. Analogously one can get lower bounds for Γ. This will be illustrated by the following numerical examples.

4. Examples.

I. We consider the following example of a biological model, having a certain interest for the cancer research. One asks for the concentration of Oxygen entering into a cell. At first we consider the simpler model in the x-y-plane. The concentration $c(0,y)=g(y)$ is given at the entering Γ_1, ($x=o$, $|y| \leq 1$), for instance

(4.1) $\quad c(o,y) = \frac{1}{2}(1-y^2)^2$ for $|y| \leq 1$ (fig.7).

The free boundary Γ is described by the unknown function

$$x = \psi(y) \qquad \text{for } |y| \leq 1,$$

$\psi(y) > 0$ for $|y| < 1$, $\psi(\pm 1) = 0$. Along Γ we have
$c = \frac{\partial c}{\partial n} = 0$ or equivalent $c = \frac{\partial c}{\partial x} = 0$ (n means the outer normal); c(x,y) satisfies the differential equation

$Lc = -\Delta c = f = -s$ with $s \geq 0$, (with $f \leq 0$) or

$$(4.2) \qquad \Delta c = s(x,y) \quad \text{in } B \quad (0<x<\psi(y),\ |y|<1).$$

s(x,y) is given, in the numerical example by $s = 20$, fig.7.
As domain B in Nr. 3 we choose $0<x$, $|y|<1$.

We look for an approximate solution w(x,y) of the form

$$(4.3) \qquad c \approx w(x,y) = \frac{1}{2}\left(1-y^2-\alpha x - \beta x^2\right)^2 (1-\gamma x).$$

We determine the parameters α, β, γ in such a way, that

$$(4.4) \qquad \Delta w \geq s.$$

c(o,y) has the prescribed boundary values (3.1) for arbitrary α,β,γ; c and c_x vanish for $x = \hat{\psi}(y)$, where $\hat{\psi}$ has to be determined from

$$1 - y^2 - \alpha x - \beta x^2 = 0 .$$

For $\beta > 0$ this is an ellipse G, one has to choose γ small enough, so that $1-\gamma x$ does not vanish in the interior of the ellipse G.
If this is satisfied, we have $-\Delta w \leq -r$ and $\hat{\psi}(y) \leq \psi(y)$ and we have got a lower bound for the free boundary. For getting good numerical results we determine the constants α,β,γ from the optimization problem

$$0 \leq \Delta w - s \leq Q, \quad Q = \text{Min}.$$

Or better from $\hat{\psi}(o)$ = Max under the restriction $o \leq \Delta u - s$.

In the same way we can determine with other values $\tilde{\alpha}, \tilde{\beta}, \tilde{\gamma}$ of the constants an upper bound $\tilde{\psi}(y)$ for the free boundary: $\tilde{\psi}(y) \geq \psi(y)$.

We get for the interesting value $\hat{x}$ with $c(\hat{x},0) = 0$ the inclusion

$$|\hat{x} - 0.2209| \leq 0.0009$$

<u>Example II:</u>

We consider the corresponding problem in the xyz = space with $r^2 = y^2 + z^2$. The entering Γ_1 may be the circle $r \leq 1, x=0$, Fig.8

The Concentration of Oxygen $c(x,y,z)$ satisfies

$$(4.5) \qquad \begin{cases} \Delta c = h \text{ in } B \\ c = g(r) \text{ on } \Gamma_1 \\ \left(\text{or } c = \frac{\partial c}{\partial r} = 0\right) \end{cases} \qquad r^2 = y^2 + cz^2$$

Free boundary Γ_2: $c = \frac{\partial c}{\partial n} = 0$

We take as Example $h = 2o$, $g(r) = \frac{1}{2}(1-r^2)^2$

Analogously to (4.4) we choose as approximate solution $v(x,r)$: $c \approx v = \frac{1}{2}(1-r^2-\alpha x-\beta x^2)^2(1-\gamma x); \alpha>o, \beta>o$.

We have on the Surface Γ^*: $v = \frac{\partial v}{\partial x} = 0$,

and we get as Domain B^*: $x>0$, $1-\gamma x>0$, $\varphi>0$ with $\varphi = 1-r^2-\alpha x-\beta x^2$;

then the comparison theorem of Nr.3 gives $B \subset B^*$ under the conditions: $-\Delta v \geq h$ in B^*

$$v \geq g \text{ on } \Gamma_1 .$$

We calculate: [$v \geq g$ is satisfied]

$$\Delta v = \left[\left(\frac{\partial \varphi}{\partial x}\right)^2 + 4r^2 - 2\varphi(\beta+2) \right] (1-\gamma x) - 2\gamma\varphi \frac{\partial \varphi}{\partial x} .$$

Δv is linear in r^2; if h is also linear in r^2, [this is true: h = const.] one has to check

$$-\Delta v \geq -h \text{ only on } S_1: r = 0,\ 0 < x < \hat{x}$$

$$\text{and on } S_2: \varphi = 0,\ 0 < x < \hat{x}$$

(Simultaneous Approximation, Bredendiek, 1969, 1970, 1976)

We get
$$\hat{x} = \begin{cases} \frac{1}{\alpha} & \text{for } \beta = 0 \\ \frac{1}{2\beta}\left[-\alpha + \sqrt{\alpha^2+4\beta} \right] & \text{for } \beta > 0. \end{cases}$$

We have, calculating with only one parameter α, $\beta = \gamma = 0$, the rough inclusion

$$\sqrt{\frac{1}{24}} \approx 0.2041 < \hat{x} < \frac{1}{4} = 0.25 .$$

We get the better inclusion with three Parameter α, β, γ :

0.21661	x	0.21927

Fig.9 illustrates the inclusion (the strip)

Strip for the Free boundary.

Inclusions for many other types of free boundary value problems have been calculated on computers in the same way.

It is easy to improve the bounds using more parameters, f.i. $v = \frac{1}{2}(1-r^2)-\alpha x-\beta x^2-\dots+\gamma x r^2+\dots)^2\,(1-\delta x-\varepsilon x^2\dots)$

(About monotonicity compare Collatz [52] [78], Collatz-Wetterling [75], Collatz-Günther-Sprekels [76] a.o.)

I thank Mr. Uwe Grothkopf for numerical calculations on a TR 4-computer.

TABLE

<table>
<tr><th></th><th></th><th>Conditions for domain</th><th>boundary</th><th>Class of functions</th></tr>
<tr><td>I.</td><td>Free boundary problem for G</td><td>$Lu = f$ in G
$u > 0$ in G
$u = 0$ outside</td><td>$u=0$
$u_n=0$
on Γ</td><td>$u \in C^2(G)$</td></tr>
<tr><td>II.</td><td rowspan="2">Boundary value problem for fixed domain B</td><td>$Lu \geq f$ in B
$u \geq 0$ in B</td><td rowspan="3">$u = k(x)$ on Γ_1</td><td rowspan="3">$u \in C'(B)$</td></tr>
<tr><td>III.</td><td>$u\,(Lu-f) = 0$
$Lu \geq f$
$u \geq 0$
} in B</td></tr>
<tr><td>IV.</td><td>Variational problem</td><td>$\int_B u(Lu-f)dx = \text{Min}$
$Lu \geq f$
$u \geq 0$
} in B</td></tr>
</table>

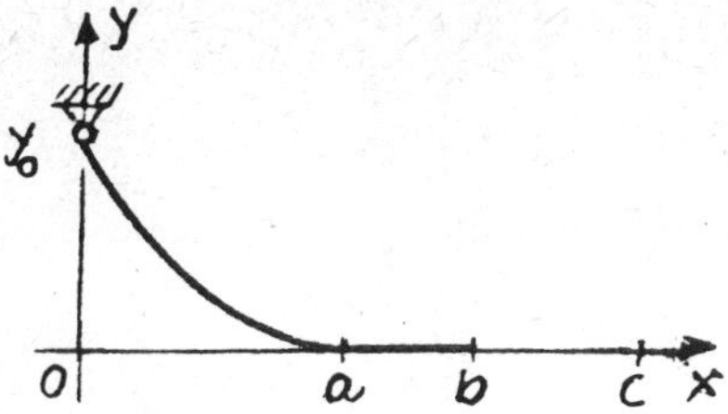

Fig. 1

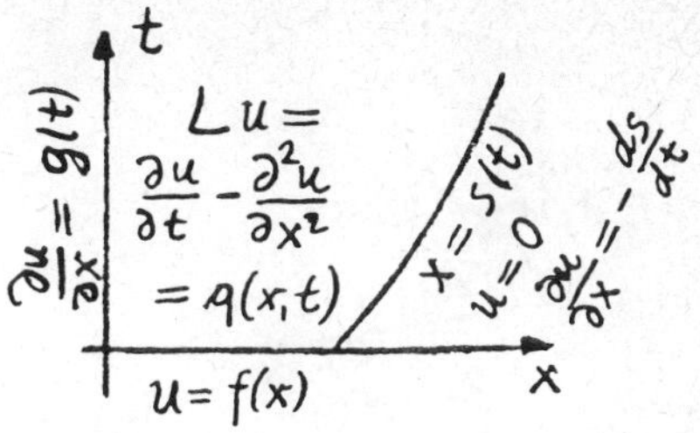

Fig. 2

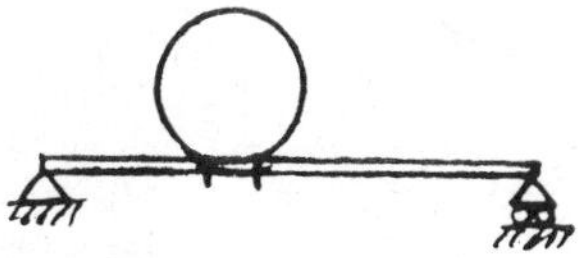

Fig. 3

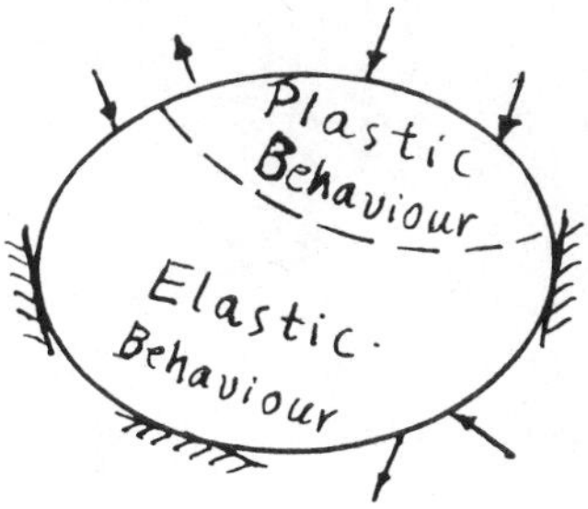

Fig. 4

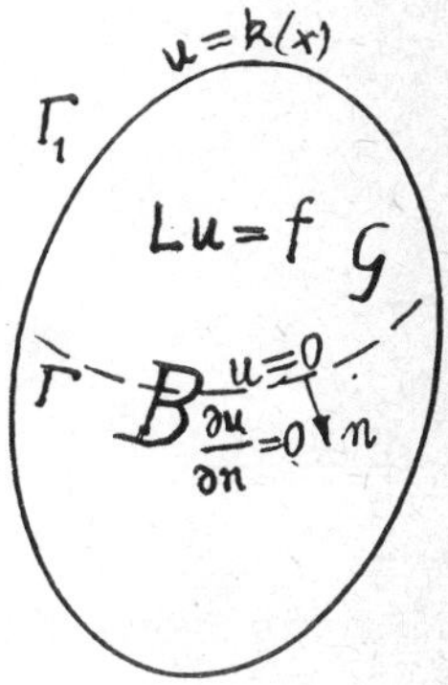

Fig. 5

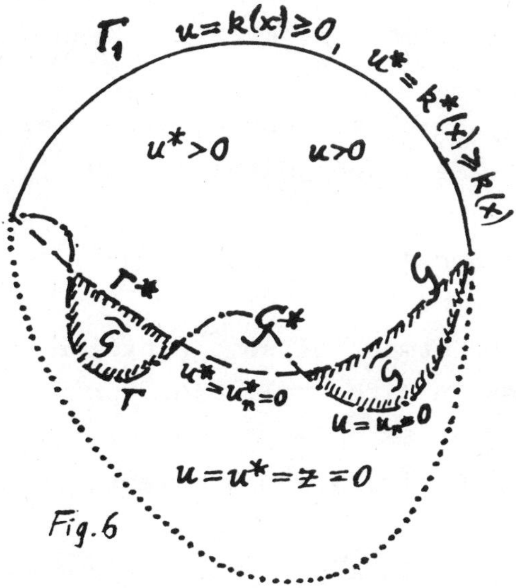

Fig. 6

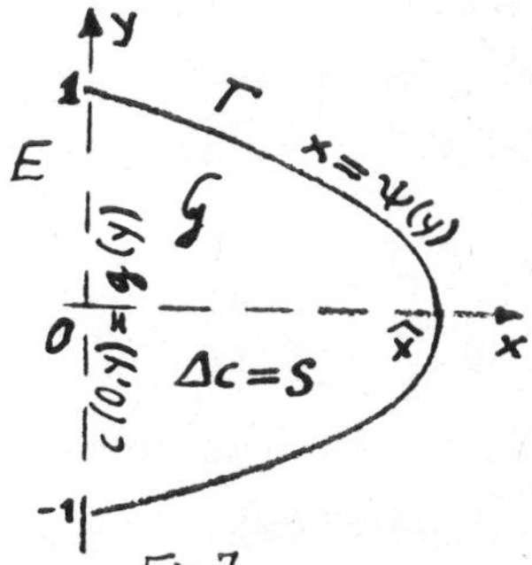

Fig. 7

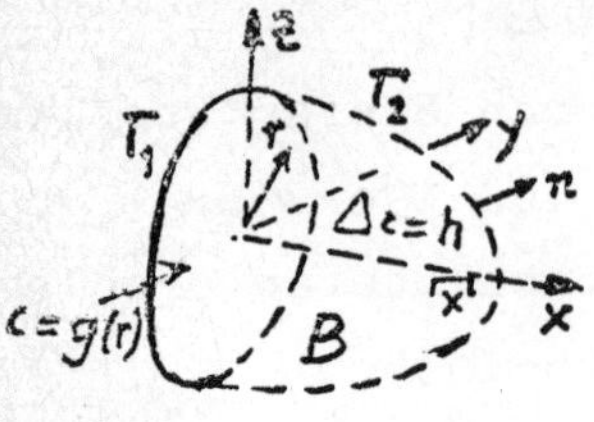

Fig.8

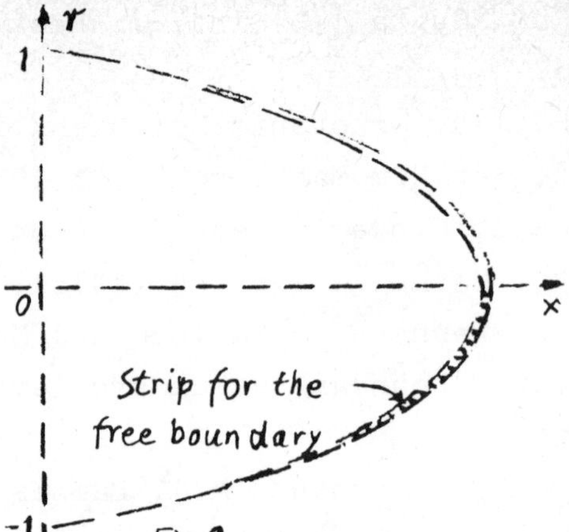

Fig. 9

References

Baiocchi, C. [72] Su un problema a frontiera libera connesso a questioni di idranlica; Ann. Pura Appl. (4) 92 (1972) 1o7-127

Bredendiek, E. [69] Simultan Approximation, Arch.Rat.Mech.Anal. 33 (1969), 3o7-33o.

Bredendiek, E. - L. Collatz [76] Simultan Approximation bei Randwertaufgaben, Internat. Ser. Num. Math. 3o (1976), 147-174

Collatz, L. [52] Aufgaben monotoner Art, Arch.Math.Anal.Mech. 3 (1952) 366-376

Collatz, L. [78] The numerical treatment of some singular boundar value problems, Lect. Notes in Math. vol. 63o, Springer 1978, 41-5o

Collatz, L. - W. Wetterling [75] Optimization problems, Springer 1975, 356 p.

Collatz, L. - H. Günther - J. Sprekels [76] Vergleich zwischen Diskretisierungsverfahen und parametrischen Methoden an einfachen Testbeispielen, Z. Angew. Math. Mech., 56 (1976), 1-11

Crank, J. - R.S. Gupta [75] Int. J. Heat Mass Transfer 18 (1975) 11o1-11o7

Hoffmann, K.H. [78] Monotonie bei nichtlinearen Stefan Problemen Internat. Ser.Num.Math. vol. 39 (1978) 162-19o

Friedman, A. - D. Kinderlehrer [75] A one phase Stefan problem, Indiana U. Math. J., vol 24 (1975), 1oo5-1o35

Meyn, K.H. - B. Werner [79]Macimum and Monotonicity Principles for elliptic boundary value problems in partioned domains, to appear

Miller, J.V. - K.W. Morton - M.J. Baines [78] A finite Element Moving Boundary Computation with an Adaptive Mesh, J. Inst. Maths Applies (1978) 22, 467-477

Ockendon, J.R. [78] Numerical and Analytic Solutions of Moving Boundary Problems, (In the book of Wilson a.o., see below, p. 129-145.

Rubinstein, L.I. [71] The Stefan Problem, Translat. Math. Monographs vol. 27, Amer. Math. Soc. 1971

Wilson, D.G. - A.D. Solomon - P.T. Boggs [78] Moving Boundary Problems Acad. Press 1978, 329 p.

COMPUTING EIGENVECTORS (AND EIGENVALUES) OF LARGE, SYMMETRIC MATRICES USING LANCZOS TRIDIAGONALIZATION

Jane Cullum

Ralph A. Willoughby

1. INTRODUCTION

Earlier reports Cullum and Willoughby [1], [2], [3] describe a Lanczos tridiagonalization procedure with no reorthogonalization for computing eigenvalues of large, symmetric matrices A in user-specified intervals. This procedure uses the Lanczos recursions, Paige [4], [5], [6],

$$\beta_{i+1}v_{i+1} = Av_i - \alpha_i v_i - \beta_i v_{i-1} \tag{1}$$

$$\alpha_i = v_i^T(Av_i - \beta_i v_{i-1}) \tag{2}$$

$$\beta_{i+1} = \| Av_i - \beta_i v_{i-1} - \alpha_i v_i \| \tag{3}$$

to generate symmetric tridiagonal matrices T_m of order m with

$$T_m(i,i) = \alpha_i \text{ , and } T_m(i,i+1) = \beta_{i+1}. \tag{4}$$

In (1) to (3), $v_0 = 0$ and v_1 is a randomly generated unit vector.

The computation of eigenvalues of A is replaced by the computation of eigenvalues of T_m and the subsequent selection of a subset of these eigenvalues as approximate eigenvalues of A. The order m required depends upon the distribution of the eigenvalues in the given matrix A. More specifically,

$$\text{Gapmax} \equiv \max_j (\lambda_j - \lambda_{j-1}) / \min_j (\lambda_j - \lambda_{j-1}) \tag{5}$$

and the local clustering of the desired eigenvalues affect the choice of m. For more details on the choice of m see Cullum and Willoughby [7]. In the eigenvalue computations in [1] and [2] only the 2 most recent Lanczos vectors were retained at each stage. If however, we save all of the Lanczos vectors $V_m = (v_1, v_2, \ldots, v_m)$ off-line as they are generated, they can, as we will see below, be used to compute associated eigenvectors for the eigenvalues obtained. Alternatively these vectors can be regenerated for the eigenvector computations.

There are 2 possible approaches to computing these eigenvectors: (1) Inverse iteration directly on $(A-\mu I)$ to generate an eigenvector x of A corresponding to μ; and (2) Inverse iteration on $(T_m-\mu I)$ to generate an eigenvector y of T_m corresponding to μ with the subsequent computation of the Ritz vector $z = V_m y$.

In Section 2 we briefly review the Paige and Saunders Lanczos algorithm SYMMLQ [8] for solving indefinite systems of equations

$$Ax = b. \tag{6}$$

In Section 3 we specialize their algorithm to inverse iteration on $(A-\mu I)$, solving

$$(A-\mu I)x = v_1. \tag{7}$$

We derive a very interesting relationship between the vector generated by SYMMLQ in solving (7) and the Ritz vectors obtained using eigenvectors of T_m. Arguments and examples demonstrate that given eigenvalues, μ_j, $1 \leq j \leq q$, of A obtained using the procedure in [1], [2], associated eigenvectors of A can be computed using SYMMLQ with the Lanczos vectors V_m generated in the eigenvalue computations. Detailed numerical results for a small matrix of order 100 are presented and demonstrate the convergence achievable.

In Section 4 we discuss and present numerical results for this matrix using the alternative approach based upon the considerably less expensive inverse iteration on $T_{m(\mu)}-\mu I$ and subsequent computation of Ritz vectors. Several problems associated with this approach are discussed and resolutions proposed. The order $m(\mu)$ used is a function of μ. In Section 5 we present the results of experiments on a diagonally-disordered matrix of order $n=1089$, Kirkpatrick [9]. Eigenvectors are computed by both approaches for various eigenvalues scattered throughout the spectrum, and the associated error estimates,

$$\|Ax-\mu x\| \text{ with } \|x\| = 1, \tag{8}$$

in these vectors are recorded. These results clearly demonstrate the effectiveness of this procedure, and in fact, that the Ritz vector approach is both economical and practical.

2. USING LANCZOS TO SOLVE SYSTEMS OF EQUATIONS

As described in Paige and Saunders [8], Lanczos tridiagonalization can be used for solving indefinite systems of equations (6). The Lanczos recursions (1) to (3) are used to generate tridiagonal T_k with $\beta_1 v_1 = b$ and $\|v_1\| = 1$. The solution y_k of

$$T_k y_k = \beta_1 e_1. \tag{9}$$

is used to define $x_k = V_k y_k$. Then from (9) and (1)

$$\|Ax_k - b\| = |\beta_{k+1} e_k^T y_k| \equiv \rho_k. \tag{10}$$

We use e_k to denote the kth coordinate vector in Euclidean m-space. The SYMMLQ algorithm requires that $\rho_k \to 0$ as $k \uparrow \infty$, although no proof of convergence is given in [8]. For details of the implementation of the above, see [8]. Using results obtained in Cullum and Willoughby [3], we have the following convergence Theorem for positive definite matrices. See [3] for the precise statements of the constructions required. Let $\beta_{2,k+1} = \prod_2^{k+1} \beta_j$.

Theorem 1. Let A be positive definite and symmetric. Generate Lanczos vectors using recursions (1) to (3) with $\beta_1 v_1 = b$ and $\|v_1\| = 1$. Using the correspondence in [3], construct associated conjugate gradient iterates w_k and residuals $r_k = -Aw_k + b$, $k = 1,2,\ldots$. Then ρ_k in (10) satisfies $\rho_k = \beta_1 \|r_{k+1}\|$. Furthermore, if $\rho_k/\rho_j \leq R$ for $j \leq k$, T_k is positive definite for all k, and the Lanczos vectors are sufficiently locally orthonormal, then $\rho_k \to 0$ as $k \uparrow \infty$.

Proof. By construction ρ_k is a multiple of the (k,1) entry in T_k^{-1}. In particular,

$$\rho_k = |\beta_{k+1} e_k^T T_k^{-1} \beta_1 e_1|.$$

Therefore, since T_k is tridiagonal,

$$\rho_k = |\beta_{1,k+1}/a_k|$$

where a_k is the determinant of T_k. But from [3] we have that

$$\beta_{2,k+1} = a_k \|r_{k+1}\|$$

Q. E. D.

We note at this point that Kaplan and Gray [10] have used (10) to compute entries in A^{-1}. To accomplish this take $v_1 = e_r$ and then multiply the vector equation corresponding to (10) first by A^{-1} and then by e_s^T. This yields the (s,r) entry in A^{-1}, if the error term $\rho_{k+1} \to 0$. In attempts to extend this convergence result to indefinite matrices we have obtained only the following weak result. We use $sp\{V\}$ to denote the space spanned by the columns of V.

Theorem 2. Let A be a symmetric nonsingular matrix. Let V_k denote the Lanczos vectors generated using A with $\beta_1 v_1 = b$ and $\|v_1\| = 1$. Let $\overline{V}_k$ be the Lanczos vectors generated using A^2 with $\overline{\beta}_1 \overline{v}_1 = Ab$ and $\|\overline{v}_1\| = 1$, and assume that the hypotheses of Theorem 1 are satisfied. Then for all $K \geq 2k$

$$sp\{V_K\} \supseteq sp\{\overline{V}_k\}. \tag{11}$$

Furthermore, given any $\varepsilon>0$, for large K there exist y_K such that with $\bar{\varepsilon} = \|A^{-1}\|\varepsilon$,

$$\|V_K T_K y_K - b\| < \bar{\varepsilon} \text{ and } \|AV_K y_K - b\| < \bar{\varepsilon}. \tag{12}$$

Comments. Observe that Theorem 2 does not state that $T_K y_K = \beta_1 e_1$, the construction given in Paige and Saunders [8], but only the weak result (12). It at least tells us that there is a good approximate solution to $Ax = b$ in $sp\{V_K\}$ for large enough K.

Proof. Observe that $sp\{V_{2k}\} = sp\{b, Ab, A^2b, \ldots, A^{2k-1}b\}$ and $sp\{\bar{V}_k\} = sp\{Ab, A^3b, \ldots, A^{2k-1}b\}$ so that (11) is clear. We know from Theorem 1 that given any $\varepsilon > 0$ for all large k,

$$\|A^2\bar{V}_k\bar{y}_k - Ab\| \le \rho_{k+1} < \varepsilon. \tag{13}$$

Choose K large enough so that

$$\bar{V}_k\bar{y}_k = V_K y_K \text{ with } y_K(K) = 0. \tag{14}$$

Then by (1), (13) and (14)

$$\|AV_K y_K - b\| = \|V_K T_K y_K - b\| = \|A\bar{V}_k\bar{y}_k - b\| < \bar{\varepsilon}.$$ Q.E.D.

The implementation in [8] does not save the Lanczos vectors. It is modelled after conjugate gradients with 2 vectors being updated at each iteration. It costs $7n$ arithmetic operations per iteration plus the cost of generating the Lanczos vector. Storage requirements are small, as low as 5 vectors of length n. An alternative implementation of Lanczos for solving system (6) being considered by Parlett [11] saves the Lanczos vectors off-line and incorporates selective reorthogonalization of the Lanczos vectors, Parlett and Scott [12].

3. USING INVERSE ITERATION ON $A-\mu I$.

From Paige [4] we know that once a 'good' eigenvalue of T_m converges then it will appear in T_k for all $k \ge m$. Too large a choice of m results in unnecessary computation, but not in less accurate eigenvalue approximations. The convergence of the eigenvalues of A, as described in [1] [2] is measured by estimates of

$$\|AVy - \mu Vy\| \text{ where } Ty = \mu y \text{ and } \|y\| = 1. \tag{15}$$

The subscript m has been dropped for simplicity. In [1], [2] these estimates were computed several different ways, but the Ritz vectors Vy were never computed explicitly. If we divide (15) by $\|Vy\|$

and by the

$$\text{Mingap}(j) \equiv \min\,(\lambda_{j+1}-\lambda_j,\ \lambda_j-\lambda_{j-1}), \tag{16}$$

then we obtain an estimate of the deviation of the Ritz vector $x = Vy/\|Vy\|$ from an eigenvector of A corresponding to λ_j closest to μ. In (16) and elsewhere, λ_j denotes an eigenvalue of A, $\lambda_1 \le \lambda_2 \le \ldots \le \lambda_n$, and we are always assuming that $\mu = \lambda_j + \varepsilon$ for some j and small ε. From Paige [5] we know that if μ is an isolated eigenvalue of T_m, then $\|Vy\|$ is not small so that (15) with (16) tells us that in this situation the Ritz vector Vy is a good approximate eigenvector of A.

As we will see in Section 4, the goodness of the Ritz vector is sensitive to the choice of m. However, appropriate choices for $m(\lambda_j)$ correspond to a large range of values bigger than the first m for which λ_j appears accurately as an eigenvalue of T_m. This question is discussed in Section 4 where Ritz vectors are used as approximate eigenvectors.

In this section we consider generating approximate eigenvectors by inverse iteration directly on $(A-\mu I)$, solving equation (7). This inverse iteration for various eigenvalues $\mu = \lambda_j + \varepsilon$ can, in fact, be accomplished using the Lanczos vectors that were used to compute μ. To see this consider the following Lemma.

<u>Lemma 1</u>. Use the Lanczos procedure in [1], [2]. Let $\mu_{k(j)} = \lambda_j + \varepsilon_j$, for j in some index set J, be eigenvalues of T_m that are good approximations to eigenvalues of A. Assume that each Mingap(j) is not 'too small'. Then approximate eigenvectors x_j of A can be obtained by solving (7) with the starting vector v_1 used in the eigenvalue computation.

Lemma 1 follows from the fact that the success of inverse iteration depends primarily upon respectable Mingaps for the eigenvalues of A being considered, and upon the desired eigenvector having a reasonable projection on v_1. SYMMLQ can be used for these computations. Furthermore, since A and $A-\mu I$ generate the same set of Lanczos vectors when the same starting vector is used, the Lanczos vectors used in the eigenvalue computations can be used to do the first iteration of inverse iteration for any of the μ. Typically only one iteration of inverse iteration is required and in fact that is all we allow the procedure to do. Thus, sets of eigenvalues can be handled simultaneously with one pass through the Lanczos vectors.

We modified SYMMLQ [8] to use the matrix T_m and the Lanczos vectors V_m generated in the eigenvalue computations. It can either read in these vectors from off-line storage or regenerate them as needed. SYMMLQ solves (9) implicitly, y_k is not generated. An LQ factorization of T_m is employed, and eigenvector iterates x_k are generated directly. In fact for each eigenvalue 2 iterates, x_k and $\bar{x}_k$, are generated and convergence occurs when the norm of either one of the corresponding residuals, r_k and $\bar{r}_k$, is less than

$$(RTOL) \|A\| \|x\| \tag{17}$$

$\|A\|$ is estimated from the α_i and RTOL is a user-specified, scale-invariant tolerance.

In each test run, convergence occurred because $\|\bar{r}_k\|$ was small. From [8], $\|\bar{r}_k\| = |\beta_{k+1} e_k^T y_k|$, the Lanczos error. Tests using a modified version of BISECT from EISPAK [13] confirmed that this convergence coincided with the initial convergence of an eigenvalue μ of T_m to the λ_j in question. The accuracy achieved for the A eigenvector depends upon the parameter RTOL. However, setting RTOL excessively small results in extra computation without corresponding significant improvements in accuracy. Thus, the work required for computing an eigenvector of A corresponding to $\mu = \lambda_j + \varepsilon$ depends upon the size of $m = m(\lambda_j)$ required for an eigenvalue of T_m to be an accurate approximation to λ_j. If Mingap(j) (see (16)) is relatively large, then $m(\lambda_j)$ is relatively small, and we can get a good approximate eigenvector of A for λ_j by using relatively few Lanczos vectors. We note that (see Table 1) $m(\lambda_j)$ can be estimated a priori, so that the user has a good estimate of the amount of computation required by SYMMLQ.

In these computations there are tradeoffs between storage requirements and the time required for the computations. Input-output calls may be expensive and it may be cheaper to regenerate the Lanczos vectors than to store them off-line and to recall them as needed. In either case SYMMLQ can be rewritten so that many eigenvalues can be considered simultaneously.

To demonstrate the convergence achievable, we ran extensive tests on a small diagonally-disordered matrix of order n=100 which we denote by KIRK100. See [2] for a description. All the eigenvalues of A were computed using the Lanczos procedure in [1], [2]. Then a subset of 60 of these eigenvalues was selected as being representative of position in the spectrum, size of Mingaps, etc. and SYMMLQ was applied to this set using the Lanczos vectors generated in the eigenvalue computations. RTOL was set to 10^{-12}. Smaller values of RTOL were also tried, but did not yield significant improvement.

Some of the numerical results are given in Table 1, see [14] for full results. The computed eigenvalues considered are listed with their Mingaps in A. ITNS is the number of Lanczos vectors used by SYMMLQ to get a corresponding eigenvector x_j and AERROR equals $\|Ax_j - \mu_j x_j\|$. AERROR/AMINGAP measures the orthogonality of the computed eigenvector w.r.t. to the eigenvector of the closest eigenvalue in A. CG = 1 means that SYMMLQ terminated because the Lanczos error ρ_k in (10) became small. The value M is an apriori estimate of the number of iterations required for convergence. For KIRK100 the eigenvalues range from -3.25 to 102.25 and the Mingaps range from 8 x 10^{-7} to .316. However, there are no eigenvalues in the interval (3.25, 97.75), so the effective spread is 11.

Table 1 demonstrates the effects of gap sizes upon the number of ITNS required. Note that the errors are all very good. KIRK100 has a tight cluster at 100.02, and looser ones elsewhere.

The SYMMLQ program that we modified was obtained from Olof Widlund who in [15] uses SYMMLQ for inverse iteration on A in conjunction with Rayleigh-Ritz iterations to compute eigenvalues and eigenvectors of large symmetric matrices. That SYMMLQ program has the option of reorthogonalizing the Lanczos vector v_{i+1} w.r.t. the 2 preceding Lanczos vectors v_i and v_{i-1}. Our Lanczos vector generation does not however reorthogonalize any vectors. We use (2) which is the modified Gram-Schmidt recommended by Paige [6], and this seems to be sufficient. The number of iterations required for convergence using our Lanczos vectors and the number required using SYMMLQ with regeneration of the Lanczos vectors and local reorthogonalization were essentially the same. We note, however, that the original SYMMLQ without the local reorthogonalization (or modified Gram-Schmidt) took considerably longer on our examples than SYMMLQ with local reorthogonalization.

In the next section we discuss the alternative approach using Ritz vectors. First, however we prove that in exact arithmetic, SYMMLQ applied to (7) yields a scaled Ritz vector. The convergence criterion in [8] selects an appropriate $m(\lambda_j)$. In the following Theorem, $a_k(\mu)$ denotes the determinant of $T_k - \mu I$ and $\hat{a}_{j+1}(\mu)$ denotes the determinant of $\hat{T}_{j+1}^{k} - \mu I$ where $\hat{T}_{j+1}$ is obtained from T_k by crossing out the first j rows and columns. Let $\beta_{j,k} \doteq \prod_{j} \beta_\ell$.

<u>Theorem 3</u>. Let $\mu = \lambda_j + \varepsilon$ with ε small and let $m = m(\mu)$ be the first m such that $a_m(\mu) = 0$. Let $z_k = V_k y_k / \rho_k$ where $\rho_k = \beta_{2,k+1}/a_k(\mu)$ be the scaled approximate eigenvector of A generated on the kth iteration of SYMMLQ on equation (7). Define $g^T = (g_1, \ldots, g_m)$ by $g_k = g_m \hat{a}_{k+1}(\mu)/\beta_{k+1,m}$ and $\|g\| = 1$. Then for $m = m(\mu)$

$$\beta_{m+1}\, g_m\, z_m = V_m g. \tag{18}$$

That is the vector $\rho_m z_m$ generated by SYMMLQ at iteration $m(\mu)$ is a multiple of the Ritz vector of $T_{m(\mu)}$ corresponding to μ.

<u>Proof</u>. Let $S^k(\mu) = \text{adjoint}\,(T_k - \mu I)$. Then by Paige [4] for $r \leq t$, the (r,t) entry of $S^k(\mu)$ is

$$\beta_{r+1,t}\, a_{r-1}(\mu)\, \hat{a}_{t+1}(\mu). \tag{19}$$

TABLE 1. KIRK100, INVERSE ITERATION ON $(A - \mu I)$ USING SYMMLQ
CONC= .7, SCALE= 100., SEED= 123456789, RTOL=10^{-12}

EVALUE μ	ITNS	M	CG?	AMINGAP	AERROR	$\frac{\text{AERROR}}{\text{AMINGAP}}$
−3.2596299508961	68	75	1	0.29941	2.7×10^{-10}	9.0×10^{-10}
−2.5839798428312	99	108	1	0.10556	1.6×10^{-10}	1.5×10^{-9}
−2.0828626930308	132	146	1	0.05388	1.1×10^{-10}	2.0×10^{-9}
−1.7496968746212	149	191	1	0.03264	9.6×10^{-11}	3.0×10^{-9}
−1.5244663935637	147	182	1	0.09483	1.1×10^{-10}	1.2×10^{-9}
−1.0664412968087	161	213	1	0.01188	7.1×10^{-11}	6.0×10^{-9}
−.39441348577218	178	227	1	0.05622	5.0×10^{-10}	8.9×10^{-9}
−.01617568670124	246	273	1	0.00921	1.3×10^{-10}	1.4×10^{-8}
−.00696736979191	264	290	1	0.00394	2.8×10^{-10}	7.1×10^{-8}
−.00302201938682	276	293	1	0.00099	1.2×10^{-10}	1.3×10^{-7}
.13732380177145	182	226	1	0.13946	5.6×10^{-10}	4.0×10^{-9}
.27866623606872	178	227	1	0.03806	1.9×10^{-10}	5.1×10^{-9}
.94132821211468	166	216	1	0.08200	1.0×10^{-10}	1.2×10^{-9}
1.0390411540704	166	216	1	0.01545	7.2×10^{-11}	4.6×10^{-9}
1.2032674765803	166	212	1	0.01677	2.6×10^{-9}	1.6×10^{-7}
2.4394857985771	106	111	1	0.09897	1.1×10^{-10}	1.1×10^{-9}
2.9316697358121	76	80	1	0.23764	1.2×10^{-10}	5.0×10^{-10}
98.736635450022	72	75	1	0.12098	3.4×10^{-8}	2.8×10^{-7}
99.018831414794	91	112	1	0.00115	7.1×10^{-10}	6.2×10^{-7}
99.029915888069	100	115	1	0.00016	1.9×10^{-9}	1.2×10^{-5}
99.041048798533	82	91	1	0.01113	7.7×10^{-10}	6.9×10^{-8}
99.334088687771	74	74	1	0.29304	9.9×10^{-10}	3.4×10^{-9}
100.00846458901	127	146	1	0.00358	1.2×10^{-9}	3.4×10^{-7}
100.02000046348	239	280	1	1.5×10^{-6}	1.6×10^{-10}	1.0×10^{-4}
100.02000199858	238	278	1	8.3×10^{-7}	8.2×10^{-10}	9.8×10^{-4}
100.02000283261	238	256	1	8.3×10^{-7}	5.6×10^{-9}	6.7×10^{-3}
100.02363205174	160	177	1	0.00363	1.0×10^{-7}	2.8×10^{-5}
100.02839544975	133	159	1	0.00476	4.9×10^{-10}	1.0×10^{-7}
100.04123450930	129	155	1	0.01001	1.7×10^{-9}	1.7×10^{-7}
100.05124118648	133	145	1	0.00873	1.5×10^{-9}	1.8×10^{-7}
100.05999107964	136	170	1	2.1×10^{-5}	4.7×10^{-10}	2.5×10^{-5}
100.07875098934	107	112	1	0.01876	1.9×10^{-10}	1.0×10^{-8}
101.02002857514	91	111	1	0.00121	1.4×10^{-9}	1.2×10^{-6}
101.03007975215	93	114	1	1.8×10^{-4}	3.2×10^{-10}	1.8×10^{-6}
101.04129787259	86	91	1	0.01104	1.7×10^{-10}	1.5×10^{-8}
101.99893770202	45	51	1	0.25688	1.7×10^{-10}	6.8×10^{-10}

By construction,

$$\beta_{2,k+1} z_k = V_k S^k(\mu) e_1. \tag{20}$$

But using the spectral expansion of $(T_k - \mu I)^{-1}$, and expressing $a_k(\mu)$ as a product of eigenvalues, at m we obtain $S^m(\mu) = a'_m(\mu) x_j x_j^T$ where $\mu = \mu_j^m$. Thus, each column of $S^m(\mu)$ is a multiple of the eigenvector of T_m corresponding to μ. From (19) we have that

$$e_k^T S^m(\mu) e_1 = \beta_{2,k} \hat{a}_{k+1}(\mu).$$

Therefore,

$$\beta_{2,m} g = g_m S^m(\mu) e_1 \tag{21}$$

is a multiple of the eigenvector of T_m corresponding to μ. Combining (20) and (21), we obtain (18). Q.E.D.

4. USING RITZ VECTORS AS APPROXIMATE EIGENVECTORS

We know that if μ is an isolated eigenvalue of T_m such that $\mu = \lambda_j + \varepsilon$ for some j and small ε, then the Ritz vector $z = Vy/\|Vy\|$ with $T_m y = \mu y$ is an approximate eigenvector of A, with the error in orthogonality bounded by

$$\|Az - \mu z\| / \text{Mingap}(j). \tag{22}$$

Lemma 2. Let μ_j, $j = 1,\ldots,q$ be isolated eigenvalues of corresponding $T_{m(j)}$. Assume that the corresponding Ritz vectors $z_j \equiv V_{m(j)} y_j / \|V_{m(j)} y_j\|$ satisfy $\|Az_j - \mu_j z_j\| = \|\varepsilon_j\| < \varepsilon$. Then for $i \neq j$,

$$|z_i^T z_j| < 2\varepsilon / |\mu_i - \mu_j|. \tag{23}$$

In other words the Ritz vectors corresponding to these different eigenvalues of A (and in fact to different size T matrices) are ε -orthonormal.

Proof:

$\mu_i z_i^T z_j = z_i^T A z_j + \varepsilon_i^T z_j = \mu_j z_i^T z_j + \varepsilon_i^T z_j + \varepsilon_j^T z_i$ and (23) follows immediately. Q.E.D.

Thus, the Ritz vectors are good approximate eigenvectors if for each $\mu = \lambda_j + \varepsilon$ we choose an appropriate $m(\lambda_j)$ reflecting the convergence of μ. The SYMMLQ subroutine does this implicitly. If we work with T instead of A, we must develop a mechanism for determining $m(\lambda_j)$. Obviously it cannot be done by computing eigenvalues for various choices of m. However, this type of information can be obtained by Sturm sequencing [17] to determine the number of eigenvalues of T_m in the interval $(\mu - \varepsilon, \mu + \varepsilon)$ as we let m increase. We can continue to increment m after μ first appears to determine if and when μ replicates. The cost for each eigenvalue is less than $6m$ arithmetic operations where m is a function of μ.

Inverse iteration on $T-\mu I$ is inexpensive. If only one iteration is sufficient, as is typically the case, then not more than 12m arithmetic operations are required per eigenvalue. Here we solve

$$(T_m-\mu I)y' = w_1 \text{ and set } y = y'/\|y'\|. \tag{24}$$

An additional 2nm arithmetic operations are required to compute the corresponding Ritz vector. This is less than 1/3 of the 7mn operations required by SYMMLQ. Neither count includes the cost of generating the Lanczos vectors. However, if we consider a set of eigenvalues simultaneously, then the cost of a single pass through the Lanczos vectors can be averaged over the set of eigenvalues.

Inverse iteration performs poorly on eigenvalues that are not well separated. Therefore, we must consider the separation of the eigenvalues of T_m. The gaps in T_m, however, are not controlled by the gaps between the eigenvalues in A. Spurious eigenvalues, eigenvalues not related to A, can appear anywhere in the spectrum of T_m. If a spurious eigenvalue appears close to an eigenvalue whose eigenvector we are trying to compute, then the computed eigenvector of T_m may be inaccurate. Examples are given below. This problem should not arise if we deal directly with A. Ritz vectors were computed for the eigenvalues of KIRK100 in Table 1, and the associated errors are given in Table 2. In Table 2, AMINGAP is the minimal gap for μ w.r.t. A, AERROR = $\|AVy-\mu Vy\| / \|Vy\|$, and TERROR = $\|T_M y-\mu y\|$ with $\|y\| = 1$. The orders M used were computed a priori as the average of $m1(\mu)$, the first value of m at which μ appears as an eigenvalue of T_m to within $\varepsilon(\mu) \equiv 10^{-10}(\max(1,\mu))$ and $m2(\mu)$, the first value of m at which μ appears as a double eigenvalue to within $\varepsilon(\mu)$. For 2 eigenvalues in Table 2, $\mu_9 = -.00696737$ and $\mu_{10} = -.00302202$, the corresponding M values are the average of $m1(\mu)$ and 410, the maximal order considered in the tests. By m = 410, these eigenvalues were not multiple to within $\varepsilon(\mu)$. In all cases the Ritz vectors computed were as good or better than the approximate eigenvectors computed using SYMMLQ on $A-\mu I$. Thus, the Ritz approach which costs approximately 1/3 as much as SYMMLQ yields comparable results.

As we noted earlier, the goodness of the Ritz vectors is affected by the order M used. Table 3 illustrates the sensitivity of the error as M is varied. For Table 3, the Ritz vectors for 3 of the eigenvalues from Table 2 were computed as M was varied from $m1(\mu)$ to $m2(\mu)$. In each case there is a range of values of M over which AERROR is essentially constant. This range always included the average of $m1(\mu)$ and $m2(\mu)$ and a large fraction of the M larger than $m1(\mu)$ but smaller than $m2(\mu)$. The large error for M near $m2(\mu)$ can be attributed to the presence of a spurious eigenvalue of T_M that is close to μ. We note that it is possible for such spurious eigenvalues to appear for values of M not close to $m2(\mu)$, and this can affect the accuracy. For examples of such occurrences see Section 5.

TABLE 2. KIRK100, COMPUTING RITZ VECTORS, EPS=$10^{-10}\mu$
CONC= .7, SCALE= 100., SEED= 123456789

EVALUE μ	M	AMINGAP	AERROR	$\frac{\text{AERROR}}{\text{AMINGAP}}$	TERROR
−3.2596300	75	0.29941	9.4×10^{-11}	3.1×10^{-10}	5.2×10^{-11}
−2.5839798	108	0.10556	6.3×10^{-11}	6.0×10^{-10}	9.4×10^{-12}
−2.0828627	146	0.05388	6.5×10^{-11}	1.2×10^{-9}	1.7×10^{-11}
−1.7496969	191	0.03265	7.1×10^{-11}	2.2×10^{-9}	1.2×10^{-12}
−1.5244664	182	0.09483	7.7×10^{-11}	8.1×10^{-10}	4.3×10^{-12}
−1.0664413	213	0.01188	9.4×10^{-11}	7.9×10^{-9}	2.5×10^{-11}
−.39441349	227	0.05622	8.2×10^{-11}	1.5×10^{-9}	4.1×10^{-11}
−.01617569	273	0.00921	1.2×10^{-10}	1.3×10^{-8}	7.8×10^{-11}
−.00696737	290	0.00394	6.4×10^{-11}	1.6×10^{-8}	1.8×10^{-11}
−.00302202	293	0.00099	7.2×10^{-11}	7.2×10^{-8}	7.5×10^{-12}
.13732380	226	0.13946	9.8×10^{-11}	7.0×10^{-10}	5.2×10^{-11}
.27866624	227	0.03806	1.1×10^{-10}	3.0×10^{-9}	6.6×10^{-11}
.94132821	216	0.08200	4.3×10^{-11}	5.2×10^{-10}	2.2×10^{-11}
1.0390412	216	0.01545	1.1×10^{-10}	7.3×10^{-9}	8.0×10^{-11}
1.2032675	212	0.01677	2.0×10^{-10}	1.2×10^{-8}	1.3×10^{-10}
2.4394858	111	0.09897	8.2×10^{-11}	8.3×10^{-10}	5.3×10^{-11}
2.9316697	80	0.23764	6.7×10^{-11}	2.8×10^{-10}	6.2×10^{-12}
98.736635	75	0.12098	7.1×10^{-11}	5.9×10^{-10}	1.5×10^{-11}
99.018831	112	0.00115	9.3×10^{-11}	8.1×10^{-8}	1.9×10^{-11}
99.029916	115	0.00016	3.8×10^{-10}	2.4×10^{-6}	1.1×10^{-10}
99.041049	91	0.01113	9.9×10^{-11}	8.9×10^{-9}	2.9×10^{-11}
99.334089	74	0.29304	1.1×10^{-10}	3.7×10^{-10}	1.1×10^{-11}
100.00846	146	0.00358	8.8×10^{-11}	2.5×10^{-8}	2.3×10^{-11}
100.02000	280	1.5×10^{-6}	1.2×10^{-10}	8.0×10^{-5}	4.6×10^{-12}
100.02000	278	8.3×10^{-7}	3.0×10^{-10}	3.6×10^{-4}	2.2×10^{-11}
100.02000	256	8.3×10^{-7}	1.5×10^{-9}	1.8×10^{-3}	4.6×10^{-10}
100.02363	177	0.00363	1.4×10^{-10}	4.0×10^{-8}	4.0×10^{-11}
100.02840	159	0.00476	4.4×10^{-10}	9.2×10^{-8}	3.4×10^{-10}
100.04123	155	0.01001	2.8×10^{-10}	2.8×10^{-8}	2.0×10^{-10}
100.05124	145	0.00873	9.6×10^{-11}	1.1×10^{-8}	5.5×10^{-11}
100.05999	170	2.1×10^{-5}	1.1×10^{-10}	5.1×10^{-6}	2.1×10^{-11}
100.07875	112	0.01876	7.9×10^{-11}	4.2×10^{-9}	8.6×10^{-12}
101.02003	111	0.001211	2.3×10^{-10}	1.9×10^{-7}	6.1×10^{-11}
101.03008	114	1.8×10^{-4}	1.7×10^{-10}	9.6×10^{-7}	1.4×10^{-10}
101.04130	91	0.01104	9.0×10^{-11}	8.1×10^{-9}	2.4×10^{-11}
101.99894	51	0.25689	8.3×10^{-11}	3.2×10^{-10}	4.7×10^{-12}

TABLE 3. KIRK100, COMPUTING RITZ VECTORS FOR VARYING M
CONC= .7, SCALE= 100., SEED= 123456789

EVALUE μ	M	AMINGAP	AERROR	$\frac{\text{AERROR}}{\text{AMINGAP}}$	$\beta_{M+1}y(M)$
−1.7496969	125	0.03265	5.2×10^{-5}	1.6×10^{-3}	5.2×10^{-5}
	145		6.5×10^{-9}	2.0×10^{-7}	6.5×10^{-9}
	155		9.1×10^{-11}	2.8×10^{-9}	4.1×10^{-11}
	165		7.1×10^{-11}	2.2×10^{-9}	3.0×10^{-11}
	175		7.1×10^{-11}	2.2×10^{-9}	7.6×10^{-12}
	185		7.1×10^{-11}	2.2×10^{-9}	1.1×10^{-12}
	195		7.1×10^{-11}	2.2×10^{-9}	5.9×10^{-12}
	205		7.1×10^{-11}	2.2×10^{-9}	1.0×10^{-13}
	215		7.1×10^{-11}	2.2×10^{-9}	3.0×10^{-13}
	225		7.1×10^{-11}	2.2×10^{-9}	2.6×10^{-12}
	235		9.2×10^{-11}	2.8×10^{-9}	5.9×10^{-11}
	258		1.4×10^{-7}	4.2×10^{-6}	1.4×10^{-7}
100.05124	92	0.00873	4.9×10^{-3}	5.6×10^{-1}	4.9×10^{-3}
	108		2.2×10^{-6}	2.5×10^{-4}	2.2×10^{-6}
	116		1.4×10^{-8}	1.6×10^{-6}	1.4×10^{-8}
	124		4.8×10^{-9}	5.5×10^{-7}	4.8×10^{-9}
	132		6.8×10^{-10}	7.8×10^{-8}	6.8×10^{-10}
	140		9.9×10^{-11}	1.1×10^{-8}	3.6×10^{-11}
	148		9.4×10^{-11}	1.1×10^{-8}	5.4×10^{-12}
	156		9.3×10^{-11}	1.1×10^{-8}	7.8×10^{-12}
	164		1.1×10^{-10}	1.3×10^{-8}	4.7×10^{-11}
	172		2.8×10^{-10}	3.2×10^{-8}	2.7×10^{-10}
	180		1.3×10^{-9}	1.5×10^{-7}	1.3×10^{-9}
	199		4.6×10^{-7}	5.3×10^{-5}	4.6×10^{-7}
100.05997	106	2.2×10^{-5}	1.2×10^{-5}	5.7×10^{-1}	1.2×10^{-5}
	124		1.2×10^{-8}	5.6×10^{-4}	1.2×10^{-8}
	133		2.1×10^{-10}	9.6×10^{-6}	1.7×10^{-10}
	142		1.3×10^{-10}	6.3×10^{-6}	8.5×10^{-11}
	151		9.7×10^{-11}	4.6×10^{-6}	1.2×10^{-11}
	160		9.7×10^{-11}	4.6×10^{-6}	1.4×10^{-12}
	169		9.9×10^{-11}	4.7×10^{-6}	1.8×10^{-12}
	178		1.9×10^{-10}	9.2×10^{-6}	1.6×10^{-10}
	187		5.5×10^{-10}	2.6×10^{-5}	5.5×10^{-10}
	196		1.1×10^{-10}	5.4×10^{-6}	4.3×10^{-11}
	205		1.1×10^{-10}	5.0×10^{-6}	9.0×10^{-12}
	235		4.2×10^{-7}	2.0×10^{-2}	4.2×10^{-7}

In Table 3 we have included the quantities $\beta_{M+1}y(M)$, which by the Lanczos formula (1) equal $\|AVy-\mu Vy\|$ to within roundoff errors. ($y(M)$ is the Mth component of y.) We observe that in fact the size of this quantity, which is computed as we do the inverse iteration on T_M, is a good predictor of poor AERRORs.

As with SYMMLQ, the Ritz vectors of a set of eigenvalues μ_j, $1 \leq j \leq q$, can be computed simultaneously corresponding to one pass through the Lanczos vector generation.

Two other observations should be mentioned. First, we observed that if M is enlarged beyond $m2(\mu)$ and incremented to $m3(\mu)$, where μ has a multiplicity of 3, a similar span of values between $m2(\mu)$ and $m3(\mu)$ produces good Ritz vectors. Of course, in practice there is no reason to go beyond $m2(\mu)$. Second, for T_m a small matrix of order 50, we used an EISPAK [13] subroutine to compute all of the eigenvectors of T_m and observed the following. Eigenvectors y of T_m corresponding to simple good eigenvalues of T_m had components $y(k)$ that decayed exponentially to '0' as $k \uparrow m$. Furthermore, eigenvectors corresponding to spurious eigenvalues had components that decayed exponentially to '0' as $k \downarrow 1$.. The computed eigenvectors corresponding to numerically multiple eigenvalues of T_m had components $y(k)$ that decayed to '0' as $k \uparrow m$ and as $k \downarrow 1$.

These tables clearly demonstrate that the Ritz vectors for appropriately chosen m are good approximate eigenvectors. Other methods for monitoring the convergence of the eigenvalues in T_m (and hence of the eigenvectors) are being considered and will be discussed elsewhere.

5. A LARGE NUMERICAL EXAMPLE

To demonstrate the effectiveness of this procedure on a large matrix, we consider a diagonally-disordered matrix KIRK1089 of order $n=1089$. We first computed the eigenvalues of this matrix using the procedure described in [1], [2] with $m=3300$, approximately $3n$. At this value of m, 1036 eigenvalues of A are approximated. There are very tight clusters of eigenvalues near 0, 99, 100 and 101 with eigenvalues that are not being approximated at $m=3300$. The eigenvector computations were all made on eigenvalues that had converged to at least 10 digit accuracy. Since the eigenvalues of A were not known a priori, this accuracy was estimated following [1], [2]. One could store the associated Lanczos vectors for use in the eigenvector computations. However, we regenerated them. This matrix has eigenvalues ranging from $\lambda_1 = -3.604$ to $\lambda_{1089} = 102.435$. The computed gaps range from less than 10^{-7} to .186. There are no eigenvalues in the interval (3.6, 97.6). Approximate eigenvectors associated with eigenvalues scattered throughout the spectrum and with various size Mingaps were computed directly using SYMMLQ to do inverse iteration on $A-\mu I$. Sample results are given in Table 4. Ritz vectors were then computed for these same eigenvalues using inverse iteration on $T-\mu I$. These results are summarized in Table 5. Reference [14] contains additional numerical results.

The headings in Tables 4 and 5 are the same as those in Tables 1 and 2, respectively. The order M of T used in each Ritz computation was the average of $m1(\mu)$ and $m2(\mu)$ whenever $m2(\mu) \leq 4904$, the maximum order we allowed. However, if $m2(\mu) > 4904$, for Tables 4 and 5 we took $M = 4904$. It would have been preferable to determine $m2(\mu)$ in each case and to always set M equal to the average of $m1(\mu)$ and $m2(\mu)$.

If one examines Tables 4 and 5, one sees that in many cases the errors obtained using the Ritz vectors are very similar to those obtained using SYMMLQ. There are eigenvalues, see $\mu_1 = -3.6046$ and $\mu_{28} = 98.17837$ for example where the error obtained using the Ritz vector is considerably better than that obtained using SYMMLQ. There are other eigenvalues see $\mu_6 = -2.3008$, $\mu_8 = -1.9024$, $\mu_{12} = -1.35715$, $\mu_{20} = -.21996$ and $\mu_{21} = 0.18282$ in Table 5, for example, where the reverse is true, the SYMMLQ vector is better. For these 5 eigenvalues the corresponding values of $\beta_{M+1}y(M)$ for $M = M(\mu)$ were 4.6×10^{-9}, 9.6×10^{-7}, 6×10^{-9}, 1.2×10^{-7} and 7.7×10^{-8}. In each case, (as was true in all other cases where these quantities were not too small) these quantities accurately reflected the AERROR achieved by the Ritz vector. Thus, they can be used as a check on the accuracy before the expensive Ritz computation is performed. At $M = M(\mu)$, there are spurious eigenvalues within 1.4×10^{-5}, 3×10^{-7} and 3×10^{-6} of μ_8, μ_{20}, and μ_{21} respectively. In such a situation inverse iteration on $T-\mu I$ can return a vector which is a mixture of the eigenvector for μ and the eigenvector for its spurious counterpart.

To test the effects of nearby spurious eigenvalues the Ritz computations were repeated on a subset of the eigenvalues in Table 5, using for each μ, $M(\mu) = m1(\mu) + 3(m2(\mu)-m1(\mu))/8$. In particular for μ_8, μ_{20}, μ_{21} significant gains in accuracy were obtained. The new AERRORs obtained were respectively, 4.2×10^{-9}, 5.8×10^{-10}, and 8.6×10^{-10}. These should be compared with the corresponding poor AERRORs in Table 5, respectively of 9.6×10^{-7}, 1.3×10^{-7} and 7.7×10^{-8}. The corresponding new values of $\beta_{M+1}y(M)$ for these eigenvalues were 3.9×10^{-9}, 2.3×10^{-10} and 7.4×10^{-10}. Observe, as noted earlier, that these quantities are good predictors of the AERRORs obtained.

The other 2 eigenvalues μ_1 and μ_{12} did not have spurious eigenvalues nearby, and we note only that the corresponding AERROR (and the $\beta_{M+1}y(M)$) resulting from the new values of M were essentially the same as those given in Table 5.

6. SUMMARY

The Lanczos procedure in [1], [2] can be used effectively to compute eigenvalues of large symmetric matrices A. We have demonstrated clearly that it can also be used to compute associated eigenvectors.

TABLE 4. KIRK1089, INVERSE ITERATION ON $(A - \mu I)$ USING SYMMLQ
CONC= .7, SCALE= 100., SEED= 123456789, RTOL=10^{-12}

EVALUE μ	ITNS	M	CG?	AMINGAP	AERROR	$\frac{\text{AERROR}}{\text{AMINGAP}}$
−3.6046373264272	169	176	1	0.03216	7.3×10^{-9}	2.3×10^{-7}
−3.0483198269417	773	943	1	0.00817	3.4×10^{-10}	4.1×10^{-8}
−2.9042580798557	814	958	1	0.00392	3.3×10^{-10}	8.4×10^{-8}
−2.7834770259638	997	1176	1	0.00308	1.1×10^{-9}	3.4×10^{-7}
−2.5554346920403	1218	1468	1	0.00327	7.4×10^{-10}	2.3×10^{-7}
−2.3008227554913	1588	1991	1	0.00430	2.7×10^{-10}	6.3×10^{-8}
−2.0915135319950	1678	2050	1	0.00568	5.6×10^{-10}	9.9×10^{-8}
−1.9024592269757	1906	2503	1	0.00172	2.6×10^{-10}	1.5×10^{-7}
−1.7328250209336	1984	2642	1	0.00207	1.2×10^{-10}	5.8×10^{-8}
−1.5512151889065	2275	2900	1	0.00261	2.6×10^{-10}	9.9×10^{-8}
−1.5476645374131	2316	2905	1	0.00350	8.2×10^{-10}	2.3×10^{-7}
−1.3571523463942	2038	2686	1	0.00665	1.3×10^{-10}	1.9×10^{-8}
−1.1860976579269	2848	3327	1	0.00196	3.8×10^{-10}	1.9×10^{-7}
−1.0237748280264	2498	3172	1	0.00481	5.7×10^{-10}	1.2×10^{-7}
−.88306080081107	3273	4904	1	0.00278	1.8×10^{-9}	6.6×10^{-7}
−.73430204125967	2835	4904	1	0.00560	3.3×10^{-10}	6.0×10^{-8}
−.59786176377525	3966	4904	1	0.00267	1.6×10^{-9}	5.9×10^{-7}
−.59519365009448	3689	4904	1	0.00267	2.6×10^{-10}	9.7×10^{-8}
−.46753394772624	3140	4904	1	0.00188	1.8×10^{-10}	9.4×10^{-8}
−.21995797406578	2969	4904	1	0.00640	1.2×10^{-9}	1.8×10^{-7}
.18282035456012	3386	4904	1	4.8×10^{-4}	4.3×10^{-10}	9.0×10^{-7}
.85286617834223	2854	4904	1	0.00271	1.5×10^{-10}	5.6×10^{-8}
.99136558610232	2673	3416	1	0.00754	2.2×10^{-9}	2.9×10^{-7}
2.5308240918541	1284	1479	1	0.00287	1.2×10^{-10}	4.3×10^{-8}
2.6188204730868	1053	1312	1	0.00514	1.2×10^{-10}	2.2×10^{-8}
2.7569962729913	1015	1239	1	0.00474	1.4×10^{-10}	3.0×10^{-8}
3.0457471775621	831	1010	1	0.00199	1.3×10^{-10}	6.7×10^{-8}
98.178376032674	449	417	1	0.00117	4.1×10^{-7}	3.5×10^{-4}
98.610657009039	1149	1237	1	5.9×10^{-5}	9.8×10^{-10}	1.6×10^{-5}
99.017505917789	979	907	1	0.00274	1.1×10^{-8}	4.7×10^{-6}
99.501541075675	526	660	1	1.3×10^{-4}	6.5×10^{-10}	4.9×10^{-6}
101.02234539870	1044	1000	1	0.00211	2.1×10^{-9}	1.0×10^{-6}
101.93034137662	326	320	1	0.02918	4.0×10^{-10}	1.4×10^{-8}
102.19679419972	269	258	1	2.0×10^{-5}	1.2×10^{-9}	6.2×10^{-5}
102.25791139576	181	188	1	0.00848	6.0×10^{-10}	7.1×10^{-8}
102.43574125475	115	113	1	0.13879	1.2×10^{-9}	9.0×10^{-9}

TABLE 5. KIRK1089, COMPUTING RITZ VECTORS, EPS=$10^{-10}\mu$
CONC= .7, SCALE= 100., SEED= 123456789

EVALUEμ	M	AMINGAP	AERROR	$\frac{\text{AERROR}}{\text{AMINGAP}}$	TERROR
−3.6046373	176	0.03216	1.8×10^{-10}	5.6×10^{-9}	1.7×10^{-10}
−3.0483198	943	0.00817	4.5×10^{-10}	5.5×10^{-8}	8.8×10^{-11}
−2.9042581	958	0.00392	2.0×10^{-10}	5.1×10^{-8}	2.0×10^{-10}
−2.7834770	1176	0.00308	5.5×10^{-10}	1.8×10^{-7}	5.9×10^{-10}
−2.5554347	1468	0.00327	2.0×10^{-9}	6.2×10^{-7}	2.3×10^{-9}
−2.3008228	1991	0.00430	1.8×10^{-8}	4.2×10^{-6}	2.0×10^{-8}
−2.0915135	2050	0.00568	7.2×10^{-10}	1.3×10^{-7}	7.4×10^{-10}
−1.9024592	2503	0.00172	9.6×10^{-7}	5.6×10^{-4}	2.0×10^{-9}
−1.7328250	2642	0.00207	3.4×10^{-9}	1.6×10^{-6}	4.0×10^{-9}
−1.5512152	2900	0.00261	6.4×10^{-10}	2.5×10^{-7}	8.1×10^{-10}
−1.5476645	2905	0.00350	3.3×10^{-9}	9.3×10^{-7}	4.2×10^{-9}
−1.3571523	2686	0.00665	8.0×10^{-9}	1.2×10^{-6}	6.0×10^{-9}
−1.1860977	3327	0.00197	3.8×10^{-10}	1.9×10^{-7}	4.6×10^{-10}
−1.0237748	3172	0.00481	1.2×10^{-9}	2.5×10^{-7}	1.4×10^{-9}
−.88306080	4904	0.00278	2.2×10^{-8}	7.9×10^{-6}	2.7×10^{-8}
−.73430204	4904	0.00560	3.5×10^{-10}	6.2×10^{-8}	4.1×10^{-10}
−.59786176	4904	0.00267	6.6×10^{-10}	2.5×10^{-7}	8.1×10^{-10}
−.59519365	4904	0.00267	1.7×10^{-10}	6.5×10^{-8}	2.0×10^{-10}
−.46753395	4904	0.00188	7.0×10^{-10}	3.7×10^{-7}	6.6×10^{-10}
−.21995797	4904	0.00640	1.3×10^{-7}	2.0×10^{-5}	8.1×10^{-10}
0.1828204	4904	0.00048	7.7×10^{-8}	1.6×10^{-4}	5.5×10^{-10}
0.8528662	4904	0.00271	5.1×10^{-10}	1.9×10^{-7}	3.3×10^{-10}
0.9913656	3416	0.00754	4.5×10^{-9}	6.0×10^{-7}	5.2×10^{-9}
2.5308241	1479	0.00287	6.9×10^{-11}	2.4×10^{-8}	5.1×10^{-11}
2.6188205	1312	0.00514	1.3×10^{-9}	2.6×10^{-7}	1.5×10^{-9}
2.7569963	1239	0.00474	8.9×10^{-10}	1.9×10^{-7}	9.8×10^{-10}
3.0457472	1010	0.00199	8.3×10^{-10}	4.2×10^{-7}	9.0×10^{-10}
98.178376	417	0.00117	1.6×10^{-9}	1.4×10^{-6}	1.4×10^{-9}
98.610657	1237	5.9×10^{-5}	7.5×10^{-10}	1.3×10^{-5}	6.1×10^{-10}
99.017506	907	0.00227	1.8×10^{-8}	8.0×10^{-6}	3.5×10^{-9}
99.501541	660	0.00013	2.1×10^{-10}	1.6×10^{-6}	2.0×10^{-10}
101.02235	1000	0.00211	5.7×10^{-9}	2.7×10^{-6}	4.5×10^{-10}
101.93034	320	0.02918	1.1×10^{-9}	3.7×10^{-8}	2.1×10^{-10}
102.19679	258	2.0×10^{-5}	4.5×10^{-9}	2.2×10^{-4}	1.8×10^{-10}
102.25791	188	0.00848	2.8×10^{-10}	3.3×10^{-8}	2.4×10^{-10}
102.43574	113	0.13879	2.2×10^{-10}	1.6×10^{-9}	9.3×10^{-11}

REFERENCES

1. Jane Cullum and Ralph A. Willoughby (1979), Fast modal analysis of large, sparse but unstructured symmetric matrices, Proceedings of the 17th IEEE Conference on Decision and Control, Jan. 10-12, 1979, San Diego, Calif., 45-53.

2. Jane Cullum and Ralph A. Willoughby (1979), Lanczos and the computation in specified intervals of the spectrum of large, sparse real symmetric matrices, eds. I. Duff and G. W. Stewart, Proceedings of the Symposium on Sparse Matrix Computations, Nov. 2-3, 1978, Knoxville, Tenn., SIAM, Philadelphia, Pa.

3. Jane Cullum and Ralph A. Willoughby (1978), The Lanczos tridigonalization and the conjugate gradient algorithms with local ε-orthogonality of the Lanczos vectors, RC 7152, IBM Research, Yorktown Heights, N.Y. (submitted to J. Linear Algebra) .

4. C. C. Paige (1971), The computation of eigenvalues and eigenvectors of very large sparse matrices, Ph.D Thesis, University of London.

5. C. C. Paige (1972), Computational variants of the Lanczos method for the eigenproblem, J. Inst. Math., Appl. 10, 373-381.

6. C. C. Paige (1976), Error analysis of the Lanczos algorithm for tridiagonalizing a symmetric matrix, J. Inst. Math. Appl., 18, 341-349.

7. Jane Cullum and Ralph A. Willoughby (1979), Computing eigenvalues of large, symmetric matrices - an implementation of a Lanczos algorithm without reorthogonalization, IBM Research Report, IBM Research, Yorktown Heights, N.Y., to appear.

8. C. C. Paige and M. A. Saunders (1975), Solution of sparse indefinite systems of linear equations, SIAM J. Numer. Anal., 12, 617-619.

9. S. Kirkpatrick (1978), private communication, IBM Research, Yorktown Heights, N.Y.

10. T. Kaplan and L. J. Gray (1976), Elementary excitations in random substitutional alloys, Phys. Rev. B, 14, 3462-3470.

11. B. N Parlett (1978), A new look at the Lanczos algorithm for solving symmetric systems of linear equations, A.E.R.E. Report CSS 64, Harwell, Oxfordshire, England.

12. B. N. Parlett and D. S. Scott (1979), The Lanczos algorithm with selective orthogonalization, Math. Comp. 33, 217-238.

13. EISPAK Guide (1976), Matrix Eigensystem Routines, Lecture Notes in Computer Science, 16, B. T. Smith et al, 2nd ed. Springer-Verlag, New York.

14. Jane Cullum and R.A. Willoughby (1979), Computing eigenvectors (and eigenvalues) of large symmetric matrices using Lanczos tridiagonalization, IBM Research Report, RC 7718, IBM Research, Yorktown Heights, N.Y.

15. Daniel B. Szyld and Olof B. Widlund (1979), Applications of conjugate gradient type methods to eigenvalue calculations, to appear.

16. G. Peters and J. H. Wilkinson (1971), The calculation of specified eigenvectors by inverse iteration, Handbook for Automatic Computation, Vol. II Linear Algebra, ed. J. H. Wilkinson - C. Reinsch, Springer-Verlag, New York, 418-439.

17. Alan Jennings (1977), Matrix Computation for Engineers and Scientists, John Wiley and Sons, New York, 279-288.

Horizontal Line Analysis of the Multidimensional Porous Medium Equation: Existence, Rate of Convergence and Maximum Principles

Joseph W. Jerome

§1. Introduction

The density φ of a homogeneous gas expanding in a homogeneous porous medium Ω satisfies the initial-value, boundary-value problem

$$(1.1)\quad \begin{array}{lll} \text{(i)} & \varphi_t = \Delta(\varphi^\gamma), & \text{in } \Omega\times(0,\infty) = D, \\ \text{(ii)} & \varphi(\cdot,0) = f, & \text{in } \Omega, \\ \text{(iii)} & \varphi = 0 & \text{on } \partial\Omega\times(0,\infty), \end{array}$$

where $f \geq 0$ has compact support in $\Omega \subset \mathbb{R}^N$ and $\gamma > 1$. The free boundary in D is the boundary of the set $\{(x,t): \varphi(x,t) > 0\}$. Existence of continuous weak solutions of the pure initial-value problem in one space variable was demonstrated by Oleinik, Kalashnikov and Yui-Lin [17] and global regularity properties were derived by Aronson [2,3] and Kruzkhov [14]; Benilan [5] has proved that φ_t is locally integrable. The free boundary in this case has been investigated by Kalashnikov [12], Aronson [4] and Knerr [13]. The multidimensional pure initial-value problem was considered by Sabinina [20] who demonstrated existence of unique weak solutions; Caffarelli and Friedman [8] have recently established continuity of solutions. The initial-value, boundary-value problem can be treated by the general theory of accretive operators in L^1 (cf. Brezis and Strauss [7] and Evans [10]), though not in L^p for $p > 1$. Brezis [6] has observed accretiveness in H^{-1}.

The only fully discrete numerical analysis of the initial-value, boundary-value problem with rates of convergence is that of Rose [18], where a homogeneous Neumann boundary condition is treated and where a piecewise linear Galerkin approximation of a regularized version of (1.1) is adjoined to backward differences in time to obtain an L^2 bounded space-time rate of $(h^{\sigma(\gamma)} + \Delta t^{1/\gamma})$ in one space dimension with appropriate adjustments in higher dimensions. Additional special results are obtained in [18] when $\gamma = 2$. A more general porous medium equation is treated in [19]. Here $\sigma(\gamma) = 2[1 + (1/\gamma)]/[2 + \gamma - (1/\gamma)]$.

Our approach employs the semidiscretization method of horizontal lines directly to (1.1), rather than a regularized version of (1.1). We make the preliminary transformation $u = K(\varphi)$, where

$$(1.2) \qquad K(\lambda) = |\lambda|^{\gamma-1}\lambda, \quad \lambda \in \mathbb{R},$$

to obtain a formulation equivalent to (1.1) if $u \geq 0$. Since the pressure is given by $\varphi^{\gamma-1}$, u is seen to be the product of the mass density and the pressure and enters as a natural variable in (1.1). The equation for u is given by

$$(1.3) \qquad \begin{array}{lll} \text{(i)} & (K^{-1}u)_t = \Delta u, & \text{in } \Omega \times (0,\infty), \\ \text{(ii)} & u(\cdot,0) = Kf = u_0, & \text{in } \Omega, \\ \text{(iii)} & u = 0, & \text{on } \partial\Omega \times (0,\infty). \end{array}$$

We introduce a strengthened version of weak solution of (1.3), necessary to derive rates of convergence. The new results of this paper involve the following: (i) a new existence proof for a weak solution u satisfying (cf. Theorem 2.1) the regularity conditions

$$(1.4) \qquad u \in X = L^\infty((0,\infty);H_0^1(\Omega)) \cap L^2(0,\infty;H_0^1(\Omega)) \cap H^1(0,\infty;L^2(\Omega)) \cap L^\infty(0,\infty;L^\infty(\Omega)),$$

based on the fully implicit Euler scheme

$$(1.5) \qquad [K^{-1}(u_m) - K^{-1}(u_{m-1})]/\Delta t = \Delta u_m;$$

(ii) a weak maximum principle, $\|u_m\|_{L^\infty(\Omega)} \leq \|u_0\|_{L^\infty(\Omega)}$, for the solutions of (1.5) which implies the known essential upper bound for u of $\|u_0\|_{L^\infty(\Omega)}$; (iii) an L^2 space-time rate of convergence of $\Delta t^{\frac{1}{2}(1+\frac{1}{\gamma})}$ for a step function approximation of u defined by $\{u_m\}$. The rate for the approximation of $K^{-1}(u)$ coincides with [18].

The plan of the paper is the following. In section two we define weak solutions of (1.3) and (1.5) and analyze the latter Dirichlet problems; in particular, we discuss maximum principles here and deduce $W^{2,r}(\Omega)$ regularity. Section three is devoted to stability estimates and section four to establishing a weak solution for (1.3). Section five discusses rates of convergence (cf. Corollary 5.3). The spaces used in this paper are described in [16].

Sections two, three and four were essentially completed in the spring of 1978. Section five, however, would not have been possible without the ideas of [18], which are adapted to the present problem. Finally, we make no explicit free boundary analysis in this paper; however, we do prove that $K^{-1}u_m(x) > 0$ if $K^{-1}u_{m-1}(x) > 0$ for the nonnegative solutions of (1.5), and this permits the construction of free boundary approximations, monotone increasing when $N = 1$.

§2. Nonlinear Dirichlet Problems and Maximum Principles

Let Ω be a bounded open subset of $\mathbb{R}^N$, $N \geq 1$, satisfying several regularity properties: (i) The negative Laplacian, with domain and range

$$-\Delta\colon W^{2,r}(\Omega) \cap H_0^1(\Omega) \to L^r(\Omega) ,$$

is an isomorphism for $1 < r < \infty$. In particular, the estimate of the open mapping theorem applies to give

$$\|v\|_{W^{2,r}(\Omega)} \leq C_r \|\Delta v\|_{L^r(\Omega)} , \quad v \in W^{2,r}(\Omega) \cap H_0^1(\Omega) . \tag{2.1}$$

(ii) The Sobolev embedding theorem holds up to the boundary in the sense of [1, pp. 97,98].

(iii) If $u \in C^1(\bar{\Omega})$, then $u \in H_0^1(\Omega) \Leftrightarrow u = 0$ on $\partial\Omega$.

(iv) Integration by parts is valid; if $v \in W^{2,r}(\Omega)$, $u \in C^1(\bar{\Omega}) \cap H_0^1(\Omega)$,

$$-\int_\Omega (\Delta v)u = \int_\Omega \nabla v \cdot \nabla u .$$

Now let $D = \Omega \times (0,\infty)$ and for $T > 0$, $D_T = \Omega \times (0,T)$. In order to define the notion of weak solution of (1.3) we write

$$H(\lambda) = K^{-1}(\lambda) = |\lambda|^{\frac{1}{\gamma}} \text{ signum } \lambda . \tag{2.2}$$

Definition 2.1. Given $0 \le H(u_0) \in C_0^\infty(\Omega)$, a function u is said to be a weak solution of (1.3) if, for every $T > 0$, $u \in C[0,T;H_0^1(\Omega)]$ and

$$\int_{D_T} [\frac{\partial\psi}{\partial t} H(u) - \nabla u \cdot \nabla\psi]\,dxdt + \int_\Omega [\psi(\cdot,0)H(u_0) - \psi(\cdot,T)H(u)(\cdot,T)]\,dx = 0 \tag{2.3}$$

for every $\psi \in H^1[0,T;L^2(\Omega)] \cap L^2(0,T;H_0^1(\Omega))$.

The following existence and regularity theorem will be proved in section four.

Theorem 2.1. There is a nonnegative weak solution u, with $\|u\|_{L^\infty(D)} \le \|u_0\|_{L^\infty(\Omega)}$, in the regularity class defined by (1.4) such that u satisfies (2.3) for every $T > 0$; u is unique.

We now describe the semidiscretization. For each integer $M \ge 1$ define

$$\Delta t = \frac{1}{M}, \quad t_m = m\Delta t . \tag{2.4}$$

The semidiscretization of (1.3i) is achieved by the following recursive scheme:

$$\text{(2.5)} \quad \begin{aligned} &\text{(i)} && [H(u_m) - H(u_{m-1})]/\Delta t - \Delta u_m = 0 \text{ in } \Omega , \\ &\text{(ii)} && u_m \in H_0^1(\Omega) , \end{aligned}$$

$m = 1,2,\dots$. The sense in which we seek solutions of (2.5) initially is the usual weak sense.

<u>Definition 2.2.</u> $u_m \in H_0^1(\Omega)$ is a weak solution of (2.5) if

$$\frac{1}{\Delta t}\int_\Omega [H(u_m)-H(u_{m-1})]\omega dx + \int_\Omega \nabla u_m \cdot \nabla \omega dx = 0 \quad \text{for all} \quad \omega \in H_0^1(\Omega) . \tag{2.6}$$

<u>Proposition 2.2.</u> There is a unique weak solution u_m satisfying (2.6) for $m = 1,2,\ldots$. Each u_m is nonnegative.

Proof: Suppose solutions $0 \le u_m \in H_0^1(\Omega)$ exist for $m < k$, $k \ge 1$. Then $H(u_{k-1}) \in H^{-1}(\Omega)$ since for $\omega \in H_0^1(\Omega)$ and $\gamma_* = 2\gamma/(2\gamma-1)$ we have by a repeated application of Hölder's inequality,

$$\begin{aligned} \left|\int_\Omega H(u_{k-1})\omega\right| &\le \|u_{k-1}\|^{1/\gamma}_{L^2(\Omega)} \|\omega\|_{L^{\gamma_*}(\Omega)} \\ &\le |\Omega|^{(\gamma-1)/(2\gamma-1)} \|u_{k-1}\|^{1/\gamma}_{L^2(\Omega)} \|\omega\|_{L^2(\Omega)} \end{aligned} \tag{2.7}$$

and (2.6) may be directly identified as the zero gradient formulation associated with the minimization of the continuous convex functional

$$G(v) = \frac{1}{2}\int_\Omega |\nabla v|^2 + \frac{1}{\Delta t}\int_\Omega J(v) - \frac{1}{\Delta t}\int_\Omega H(u_{k-1})v \tag{2.8}$$

over $H_0^1(\Omega)$; here

$$J(\lambda) = \int_0^\lambda H(t)dt = \frac{\gamma}{\gamma+1}|\lambda|^{1+1/\gamma} \tag{2.9}$$

is the convex primitive of H. The continuity of $\int_\Omega J(v)$ follows since it is subquadratic; more precisely, by the mean value theorem and (2.7),

$$\begin{aligned} &\left|\int_\Omega J(v_1) - \int_\Omega J(v_2)\right| \\ &\le |\Omega|^{(\gamma-1)/(2\gamma-1)} \|\max(|v_1|,|v_2|)\|^{1/\gamma}_{L^2(\Omega)} \|v_1 - v_2\|_{L^2(\Omega)} . \end{aligned} \tag{2.10}$$

Standard results [9] guarantee the existence of a unique minimum $G(u_k)$ for G over $H_0^1(\Omega)$, since G is continuously Gâteaux differentiable, and (2.6) expresses the necessary and sufficient condition

$$[G'(u_k)](\omega) = 0 \quad \text{for all} \quad \omega \in H_0^1(\Omega) .$$

To prove $u_k \geq 0$ suppose the contrary, i.e., suppose there exists a set $\Omega_* \subset \Omega$ of positive measure such that

$$(2.11) \qquad u_k(x) < 0 , \qquad x \in \Omega_* .$$

Now select $\omega = (u_k)_-$ in (2.6), where

$$(u_k)_-(x) = \begin{cases} 0 , & u_k(x) > 0 , \\ u_k(x) , & u_k(x) \leq 0 . \end{cases}$$

It is known [21] that $(u_k)_- \in H_0^1(\Omega)$ and by elementary integration theory and (2.11),

$$(2.12) \qquad \int_{\Omega_*} |u_k|^{1+1/\gamma} > 0 .$$

Thus, from (2.6),

$$(2.13) \qquad \frac{1}{\Delta t}\int_\Omega H(u_k)(u_k)_- + \int_\Omega \nabla(u_k)\cdot\nabla(u_k)_- = \frac{1}{\Delta t}\int_\Omega H(u_{k-1})(u_k)_-$$

so that

$$\int_{\Omega_*} |u_k|^{1+1/\gamma} = \int_{\Omega_*} H(u_k)(u_k)_- \leq \int_\Omega H(u_k)(u_k)_- + \Delta t \int_\Omega |\nabla(u_k)_-|^2$$

$$= \int_\Omega H(u_{k-1})(u_k)_- \leq 0 ,$$

which contradicts (2.12). Thus, $u_k \geq 0$ and the proof is completed.

Remark 2.1. The following proposition gives regularity results and maximum type principles for the $\{u_m\}$.

Proposition 2.3. The following statements concerning solutions of (2.6) hold:

(i) $u_m \in W^{2,r}(\Omega)$, all $1 < r < \infty$, hence (2.5i) holds pointwise a.e. in Ω;

(ii):

(2.14) $$\|u_m\|_{L^\infty(\Omega)} \leq \|u_{m-1}\|_{L^\infty(\Omega)} \leq \|u_0\|_{L^\infty(\Omega)}, \quad m \geq 1 ;$$

(iii) $$u_m(x) \neq 0 \text{ if } u_{m-1}(x) \neq 0, \quad x \in \Omega .$$

Proof: Assume, inductively, that (2.14i) holds for $m < k$. We shall use a bootstrap method to verify (2.14i) for $m = k$. Thus, since $\Delta u_k \in L^{2\gamma}(\Omega)$, the solution u_k of (2.6) is in $W^{2,2\gamma}(\Omega)$ by (2.1). Setting $p_0 = 2$, we see by the Sobolev embedding theorem [1, pp. 97, 98] that $u_k \in L^{p_1}(\Omega)$ with $p_1 = n\gamma p_0/(n-2\gamma p_0)$ if $n - 2\gamma p_0 > 0$; otherwise (2.14i) is already established. Thus $H(u_k) \in L^{p_1\gamma}(\Omega)$ and again gives $u_k \in W^{2,p_1\gamma}(\Omega)$; the Sobolev embedding theorem yields $u_k \in L^{p_2}(\Omega)$ with $p_2 = n\gamma p_1/(n-2\gamma p_1)$ if $n - 2\gamma p_1 > 0$; otherwise (2.14i) is established. One sees easily that the sequence p_i terminates at an i_0 for which $n - 2\gamma p_{i_0} \geq 0$. It follows that $u_m \in W^{2,r}(\Omega)$, $1 < r < \infty$, and hence integration by parts in (2.6) for $\omega \in C_0^\infty(\Omega)$ leads to the conclusion that the equation holds pointwise. To verify (2.14ii), let ℓ be any positive integer and set $s - 1 = \ell\gamma$. Multiplying (2.5i) by $[H(u_m)]^{s-1}$ and applying the inequality

(2.15) $$\lambda\nu \leq \frac{\lambda^p}{p} + \frac{\nu^q}{q}, \quad \lambda \geq 0, \quad \nu \geq 0, \quad \frac{1}{p} + \frac{1}{q} = 1 ,$$

with $\lambda = [H(u_m)]^{s-1}$, $\nu = H(u_{m-1})$, $p = s/(s-1)$, $q = s$, yields

(2.16) $$\frac{[H(u_m)]^s}{s\Delta t} - (\Delta u_m)u_m^\ell \leq \frac{[H(u_{m-1}]^s}{s\Delta t}, \quad \text{in } \Omega .$$

Integrating (2.16) by parts over Ω gives

$$\int_\Omega [H(u_m)]^s \leq \int_\Omega [H(u_{m-1})]^s$$

since $\int_\Omega |\nabla u_m|^2 u_m^{\ell-1} \geq 0$. Taking sth. roots and letting $s \to \infty$ gives

$$(2.17) \qquad \|H(u_m)\|_{L^\infty(\Omega)} \leq \|H(u_{m-1})\|_{L^\infty(\Omega)}, \qquad m \geq 1 .$$

(2.14ii) follows from (2.17); (2.14iii) is immediate from (2.5i).

§3. Stability Estimates

Proposition 3.1. There is a positive constant C such that

$$(3.1) \quad \begin{aligned} &\text{(i)} \qquad \sum_{m=1}^{\infty} \|u_m - u_{m-1}\|^2_{L^2(\Omega)} \leq C\Delta t , \\ &\text{(ii)} \qquad \|\nabla u_m\|_{L^2(\Omega)} \leq \|\nabla u_0\|_{L^2(\Omega)}, \qquad m \geq 1 . \end{aligned}$$

Lemma 3.2. For each $m = 1, 2, \ldots$ define

$$(3.2) \qquad \tilde{u}_m(x) = \gamma^{-\frac{1}{2}} \begin{cases} 0, & \text{if } u_{m-1}(x) = u_m(x), \\ [u_m(x) - u_{m-1}(x)]/[\max(u_{m-1}(x), u_m(x))]^{(\gamma-1)/(2\gamma)}, & \text{otherwise.} \end{cases}$$

Then,

$$(3.3) \qquad \frac{1}{\Delta t} \sum_{m=1}^{\infty} \|\tilde{u}_m\|^2_{L^2(\Omega)} \leq \frac{1}{2}\|\nabla u_0\|^2_{L^2(\Omega)} .$$

Proof of Lemma 3.2: For $x \geq 0$, $y \geq 0$, $x \neq y$ and $r > 0$, $s > 0$, $r \neq s$ the inequality (cf. [15, p. 85],

$$(3.4)\qquad \left(\frac{y^s - x^s}{y^r - x^r}\,\frac{r}{s}\right)^{1/(s-r)} \le \max(x,y)$$

leads to the inequality a.e. in Ω:

$$(3.5)\qquad |u_m - u_{m-1}| \le \gamma[\max(u_{m-1}, u_m)]^{1-1/\gamma}\,|u_m^{1/\gamma} - u_{m-1}^{1/\gamma}| ,$$

if the identifications $y = u_m$, $x = u_{m-1}$, $r = 1/\gamma$ and $s = 1$ are made. Note that this substitute for the mean value theorem requires no differentiability. If (2.6) is used as a starting point with $\omega = u_m - u_{m-1}$ and $|u_m^{1/\gamma} - u_{m-1}^{1/\gamma}|$ is estimated pointwise from below by (3.5), we have

$$(3.6)\qquad \frac{1}{\Delta t}(\tilde{u}_m , \tilde{u}_m) + a(u_m , u_m - u_{m-1}) \le 0 ,$$

for $m = 1, 2, \ldots$. Here we have written

$$a(v,w) = \int_\Omega \nabla v \cdot \nabla w .$$

If the inequality

$$(3.7)\qquad a(u_m , u_m - u_{m-1}) \ge \frac{1}{2}\, a(u_m , u_m) - \frac{1}{2}\, a(u_{m-1} , u_{m-1})$$

is used in (3.6) and the resultant inequalities added over the indices $m = 1, \ldots, k$ we obtain

$$(3.8)\qquad \frac{1}{\Delta t}\sum_{m=1}^{k} \|\tilde{u}_m\|^2_{L^2(\Omega)} + \frac{1}{2}\, a(u_k , u_k) \le \frac{1}{2}\, a(u_0 , u_0) .$$

This proves (3.3) since k is arbitrary.

Proof of Proposition 3.1: To obtain (3.1ii), we simply use (3.8); the result is immediate. To obtain (3.1i), we use the estimate

$$(3.9)\qquad \sum_{m=1}^{k} \|u_m - u_{m-1}\|^2_{L^2(\Omega)} \leq \gamma \|u_0\|^{1-(1/\gamma)}_{L^\infty(\Omega)} \sum_{m=1}^{k} \|\tilde{u}_m\|^2_{L^2(\Omega)}$$

for all $k \geq 1$ in conjunction with (3.3). The proof of the proposition is completed.

Proposition 3.3. For $p = 1 + \frac{1}{\gamma}$ and $q = \gamma + 1$ the estimate

$$(3.10)\qquad \sum_{m=1}^{\infty} \int_\Omega |\nabla u_m|^2 \Delta t \leq \left(\frac{1}{q}\right) \|u_0\|^p_{L^p(\Omega)}$$

holds.

Proof: Set $\omega = u_m$ in (2.6) to obtain

$$(3.11)\qquad \int_\Omega u_m^{(1+(1/\gamma))} - \int_\Omega u_{m-1}^{1/\gamma} u_m + a(u_m, u_m)\Delta t = 0 .$$

Setting $p = 1 + 1/\gamma$ and using the inequality (2.15), with $\lambda = u_m$ and $\nu = u_m^{1/\gamma}$, we obtain from (3.11),

$$\frac{1}{q}\int_\Omega u_m^p - \frac{1}{q}\int_\Omega u_{m-1}^p + a(u_m, u_m) = 0 .$$

Summing over $m = 1, \ldots, k$ gives

$$\frac{1}{q}\|u_k\|^p_{L^p(\Omega)} + \sum_{m=1}^{k} a(u_m, u_m)\Delta t \leq \frac{1}{q}\|u_0\|^p_{L^p(\Omega)}$$

which establishes (3.10).

For the remainder of this section we shall discuss natural piecewise linear and step function sequences whose topological properties are related to the stability estimates derived earlier in this section.

Definition 3.1. The sequences $\{U^M_{PL}\}$ and $\{\Phi^M_{PL}\}$ are obtained by piecewise linear interpolation in t of the values $\{u_m\}$ and $\{H(u_m)\}$ respectively obtained for fixed $\Delta t = 1/M$. Specifically,

(3.12) $$U_{PL}^{M}(x,t) = u_m(x) + \frac{(t-m\Delta t)}{\Delta t}(u_{m+1}(x)-u_m(x))$$

for $m\Delta t \leq t \leq (m+1)\Delta t$, $m = 0,1,\ldots$ with a similar definition for Φ_{PL}^{M}. The sequences $\{U_S^M\}$ and $\{\Phi_S^M\}$ are step function sequences:

(3.13)
(i) $$U_S^M(x,t) = u_m(x), \quad m\Delta t \leq t < (m+1)\Delta t, \quad m \geq 0,$$
(ii) $$\Phi_S^M(x,t) = H(u_m)(x), \quad m\Delta t \leq t < (m+1)\Delta t, \quad m \geq 0.$$

<u>Remark 3.1</u>. One can show using (3.1i) that the weak convergence in $L^2(D)$ of $\{U_{PL}^M\}$ or $\{U_S^M\}$ implies that of the other with the same limit. Similar statements hold concerning the convergence of these sequences in $L^2(D)$ or $L^2(D_T)$.

<u>Proposition 3.4</u>. The sequence $\{U_{PL}^M\}$ is bounded in the space X defined by (1.4). In particular, there is a subsequence $\{M_i\}$ of $\{M\}$ and a function $u \in X$ such that

(3.14)
(i) $U_{PL}^{M_i} \rightharpoonup * u$, $U_S^{M_i} \rightharpoonup * u$ (weak * in $L^\infty(0,\infty;H_0^1(\Omega))$),

(ii) $U_{PL}^{M_i} \rightharpoonup u$, $U_S^{M_i} \rightharpoonup u$ (weakly in $L^2(0,\infty;H_0^1(\Omega))$),

(iii) $U_{PL}^{M_i} \rightharpoonup u$, (weakly in $H^1(0,\infty;L^2(\Omega))$),

(iv) $U_{PL}^{M_i} \rightharpoonup * u$ (weak * in $L^\infty(D)$),

(v) $U_S^{M_i}|_{D_T} \rightarrow u|_{D_T}$ (in $L^2(D_T)$ for every $T > 0$),

(vi) $U_S^{M_i}(x,t) \rightarrow u(x,t)$, a.e. in D.

Moreover, the pointwise convergence of $U_S^{M_i}$ in D implies that

(vii) $\Phi_S^{M_i}|_{D_T} \rightarrow H(u)|_{D_T}$ (in $L^1(D_T)$ for every $T > 0$.

Proof: The boundedness of $\{U^M\}$ in X follows from (3.1ii), (3.10), (3.1i) and (2.14ii) respectively. By the weak * compactness (or weak compactness) properties of the spaces defining X it follows that there is a single subsequence $\{U_{PL}^{M_i}\}$ weak * (or weakly) convergent in each of these spaces to a limit depending upon the space involved. These four limits are seen to coincide by noting that the compact injection of $H^1[0,T;L^2(\Omega)] \cap L^2(0,T;H_0^1(\Omega))$ into $L^2(D_T)$, $T > 0$, implies that $\{U_{PL}^{M_i}\}$ is convergent in $L^2(D_T)$ for every $T > 0$. Note that this follows from the compact injection of $H_0^1(\Omega)$ into $L^2(\Omega)$. By Remark 3.1, $U_S^{M_i}$ is convergent as described in (i,ii,v) to the same limit. Thus, (3.14i-v) hold and without loss of generality a diagonalization procedure, based on $T_i \to \infty$, permits us to assume (3.14vi). Now by (2.14ii),

$$\int_{D_T} U_S^{M_i} H(U_S^{M_i}) \le (1/M_i) \sum_{m=0}^{[TM_i]} \int_\Omega |u_m|^{1+1/\gamma} \le C_T .$$

Since $U_S^{M_i}$ is pointwise convergent to u on D_T, it follows that $H(U_S^{M_i})|_{D_T} = \Phi_S^{M_i}|_{D_T}$ is pointwise convergent to $H(u)|_{D_T}$ on D_T. By an L^1 convergence lemma of Strauss [22, Th. 1.1], it follows that (3.14vii) holds. This completes the proof.

§4. Convergence to the Weak Solution: Existence

We shall prove Theorem 2.1 in this section. By a denseness argument, we may assume that the test function ψ of (2.3) satisfies $\psi \in C^\infty(\bar{D}_T)$, $\psi = 0$ on $\partial\Omega \times (0,T)$. Thus, for such a function ψ define

$$\psi_m(x) = \frac{1}{\Delta t} \int_{m\Delta t}^{(m+1)\Delta t} \psi(x,t)\,dt . \tag{4.1}$$

Setting $\omega = \psi_{m-1}$ in (2.6) we find, after adding and summing by parts,

$$\sum_{m=1}^{L-2} (H(u_m), \frac{\psi_{m-1}-\psi_m}{\Delta t})_{L^2(\Omega)} \Delta t + (H(u_{L-1}),\psi_{L-2})_{L^2(\Omega)}$$

(4.2)

$$- (H(u_0),\psi_0)_{L^2(\Omega)} + \sum_{m=1}^{L-1} a(u_m,\psi_{m-1})\Delta t = 0 ,$$

where $L = [\frac{T}{\Delta t}] = [MT]$. Rewriting (4.2), we have

$$\int_{\Delta t}^{(L-1)\Delta t} (\Phi_S^M , \xi^M)_{L^2(\Omega)} dt + (H(u_{L-1}),\psi_{L-2})_{L^2(\Omega)}$$

(4.3)

$$- (H(u_0),\psi_0)_{L^2(\Omega)} + \int_{\Delta t}^{L\Delta t} a(U_S^M , \psi_{\Delta t}^M) dt = 0$$

where ψ^M is the step function defined on D and determined by the "steps" $\{\psi_m\}$ of (4.1) as in (3.13), where ξ^M is the negative of the backward difference of ψ^M, and where $\psi_{\Delta t}^M$ is the translate of ψ^M by $-\Delta t$. Now, (cf. [11]),

(i) $$\psi^M \to \psi \quad \text{in} \quad L^2(0,T;H_0^1(\Omega))$$

(4.4) and

(ii) $$\xi^M \to -\frac{\partial\psi}{\partial t} \quad \text{in} \quad L^\infty(D) .$$

Noting also that $\psi_0 \to \psi(\cdot,0)$ and $\psi_L \to \psi(\cdot,T)$ in $L^\infty(\Omega)$ as $L \to \infty$ through values $[M_i T]$, we obtain from (3.14), (4.3) and (4.4) the relation,

$$-\int_{D_T} H(u)\frac{\partial\psi}{\partial t} dxdt + \int_{\Omega\times\{T\}} \psi(\cdot,T)H(u)dx - \int_{\Omega\times\{0\}} \psi(\cdot,0)H(u_0)dx + \int_D \nabla u\cdot\nabla\psi \, dxdt ,$$

which is just (2.3).

The nonnegativity of u follows from the nonnegativity of the members of the sequence $\{U_S^M\}$ and from (3.14vi); the boundedness of u follows from (2.14ii). This proves Theorem 2.1 since the uniqueness was established in [17]; the proof there holds both for the pure initial-value and the boundary-value problem.

<u>Remark 4.1</u>. The uniqueness of u establishes that the choice $\{M_i\} = \{M\}$ holds in (3.14), i.e. the full sequences converge. Moreover, it can be shown that

$\Phi_{PL}^{M} \to H(u)$ in $L^1(D_T)$. This, however, requires the stability estimate, for $q = 1 + \frac{1}{\gamma}$,

$$\sum_{m=1}^{\infty} \|H(u_m) - H(u_{m-1})\|_{L^q(\Omega)}^q \le \frac{1}{2} \int_\Omega |\nabla u_0|^2 \Delta t \ .$$

§5. Convergence Rates

Definition 5.1. We denote by S the inverse of $-\Delta$ when the latter is viewed as a mapping of $H_0^1(\Omega)$ onto $H^{-1}(\Omega)$. Thus, $v = Sf$ if and only if

(5.1i) $$a(v,w) = \langle f,w\rangle, \quad \text{for all} \quad w \in H_0^1(\Omega) \ .$$

Thus, by (2.1), the restriction of S to $L^2(\Omega)$ satisfies

(5.1ii) $$a(Sf,w) = (f,w)_{L^2(\Omega)} \ .$$

We define a norm in $H^{-1}(\Omega)$ by

(5.1iii) $$\|f\|_{H^{-1}(\Omega)} = \langle f,Sf\rangle^{\frac{1}{2}} = [a(Sf,Sf)]^{\frac{1}{2}} \ .$$

Remark 5.1. The restriction of S to $L^2(\Omega)$ is self-adjoint and positive definite.

Proposition 5.1. Let u be the unique solution of (2.3). Then $SH(u)\big|_{D_T} \in H^2(D_T)$, $T > 0$, and the equation

(5.2) $$\frac{\partial SH(u)}{\partial t} + u = 0$$

holds a.e. in Ω for each $0 < t < T$.

Proof: For $\zeta \in C_0^\infty(D_T)$ let $\psi = S\zeta$ in (2.3). Using the commutative relation $(\frac{\partial}{\partial t} S - S\frac{\partial}{\partial t})\zeta = 0$ and the self-adjointness of S we obtain

(5.3) $$\int_{D_T} SH(u)\,\frac{\partial\zeta}{\partial t} = \int_{D_T} u\,\zeta$$

so that the distribution derivative $\frac{\partial}{\partial t} SH(u)$ is equal to $-u$. Since $H(u) \in L^2(D_T)$, it follows that $\frac{\partial^2}{\partial x_i^2} SH(u) \in L^2(D_T)$, $i = 1,\ldots,N$ and the remaining second order partials are in $L^2(D_T)$ from the properties of u. It follows that $SH(u) \in H^2(D_T)$. In particular, (5.3) can be integrated by parts to obtain (5.2), since ζ is arbitrary and the left side of (5.2) is continuous in t as a mapping into $L^2(\Omega)$.

<u>Theorem 5.2</u>. Let $d_m = u(t_m) - u_m$ and $e_m = H(u)(t_m) - H(u_m)$. Then the estimates

(5.4)

(i) $$\|e_k\|^2_{H^{-1}(\Omega)} + \frac{\gamma}{\gamma+1}\sum_{m=1}^{k}\|e_m\|^{\gamma+1}_{L^{\gamma+1}}\Delta t \le C_1\int_{D_{T_k}} \left|\frac{\partial u}{\partial t}\right|^{1+(1/\gamma)} \Delta t^{1+(1/\gamma)}, \quad T_k = k\Delta t\,,$$

(ii) $$\sum_{m=1}^{k}\|d_m\|^2_{L^2(\Omega)}\Delta t \le C_2\int_{D_{T_k}} \left|\frac{\partial u}{\partial t}\right|^{1+(1/\gamma)} \Delta t^{1+(1/\gamma)}, \quad T_k = k\Delta t\,.$$

hold for every $k \ge 1$, where C_1 and C_2 are positive constants.

Proof: Applying S to (2.5i) and setting $t = t_m$ in (5.2) gives

(i) $$S[H(u_m) - H(u_{m-1})]/\Delta t + u_m = 0\,,$$

(5.5) (ii) $$[SH(u)(t_m) - SH(u)(t_{m-1})]/\Delta t + u(t_m)$$

$$= \{[SH(u)(t_m) - SH(u)(t_{m-1})]/\Delta t - \frac{\partial SH(u)}{\partial t}(t_m)\}$$

$$= \frac{1}{\Delta t}\int_{t_{m-1}}^{t_m} [\frac{\partial SH(u)}{\partial t}(s) - \frac{\partial SH(u)}{\partial t}(t_m)]\,ds$$

$$= \frac{1}{\Delta t}\int_{t_{m-1}}^{t_m}\int_{t_m}^{s} \frac{\partial^2 SH(u)}{\partial t^2}(\tau)\,d\tau\,ds\,.$$

Subtracting (5.5i) from (5.5ii), multiplying by e_m and integrating over Ω gives

$$\frac{1}{\Delta t}\left[\int_\Omega e_m S e_m - \int_\Omega e_m S e_{m-1}\right] + \int_\Omega e_m d_m \tag{5.6}$$

$$= \frac{1}{\Delta t}\int_\Omega e_m\left\{\int_{t_{m-1}}^{t_m}\int_s^{t_m}\frac{\partial u}{\partial t}\,d\tau ds\right\}dx = \int_\Omega e_m f_m ,$$

where we have noticed that $\frac{\partial u}{\partial t} = -\frac{\partial^2}{\partial t^2}SH(u)$ and where (5.6) defined f_m. The inequality

$$y^\lambda - x^\lambda \leq (y-x)^\lambda , \quad y \geq x, \quad 0 \leq \lambda \leq 1 ,$$

with $\lambda = \frac{1}{\gamma}$, leads to $|d_m| \geq |e_m|^\gamma$ so that

$$\int_\Omega e_m d_m \geq \int_\Omega |e_m|^{\gamma+1} . \tag{5.7}$$

Also, (2.15), with $p = \gamma + 1$ and $q = 1 + \frac{1}{\gamma}$, yields

$$\int_\Omega |e_m f_m| \leq \frac{1}{\gamma+1}\int_\Omega |e_m|^{\gamma+1} + \frac{\gamma}{\gamma+1}\int_\Omega |f_m|^{1+(1/\gamma)} . \tag{5.8}$$

Using (5.7), (5.8) and the fact that the first two terms in (5.6) are bounded from below by

$$\frac{1}{2}\|e_m\|^2_{H^{-1}(\Omega)} - \frac{1}{2}\|e_{m-1}\|^2_{H^{-1}(\Omega)} ,$$

we obtain, after summing (5.6) over $m = 1,\ldots,k$:

$$\frac{1}{2}\|e_k\|^2_{H^{-1}(\Omega)} + \frac{\gamma}{\gamma+1}\sum_{m=1}^{k}\|e_m\|^{\gamma+1}_{L^{\gamma+1}(\Omega)}\Delta t \leq \frac{\gamma}{\gamma+1}\sum_{m=1}^{k}\int_\Omega |f_m|^{1+(1/\gamma)}\Delta t \tag{5.9}$$

Finally, we have by Hölder's inequality, and the limiting integral form of the triangle inequality,

$$(5.10)\quad \begin{aligned} \|f_m\|^{1+(1/\gamma)}_{L^{1+(1/\gamma)}(\Omega)} &\le \Big[\frac{1}{\Delta t}\int_{t_{m-1}}^{t_m}\int_s^{t_m}\Big\|\frac{\partial u}{\partial t}\Big\|_{L^{1+(1/\gamma)}(\Omega)} d\tau ds\Big]^{1+(1/\gamma)} \\ &\le (\Delta t)^{(1/\gamma)}\int_{t_{m-1}}^{t_m}\Big\|\frac{\partial u}{\partial t}\Big\|^{1+(1/\gamma)}_{L^{1+(1/\gamma)}(\Omega)} ds\ , \end{aligned}$$

so that (5.9) and (5.10) lead to (5.4i). To obtain (5.4ii), replace (5.7) by

$$\gamma\| u_0 \|^{1-(1/\gamma)}_{L^\infty(\Omega)} \int_\Omega e_m d_m \ge \int_\Omega d_m^2\ ,$$

which is derived from (3.4) and the maximum principle, and proceed as before. This completes the proof.

We obtain immediately the following corollary, upon using the fact that $u_t \in L^2(D)$.

Corollary 5.3. There are positive constants K_1 and K_2 such that, for every $T > 0$, the estimates

$$(5.11)\quad \begin{aligned} &\text{(i)}\quad \sup_{0\le t\le T}\|H(u)-\Phi_S^M\|_{H^{-1}(\Omega)} + \|u-U_S^M\|_{L^2(D_T)} \\ &\qquad \le K_1\|u_t\|^{(1/2)(1+(1/\gamma))}_{L^2(D)}\ T^{(1/4)(1-(1/\gamma))}\ \Delta t^{(1/2)(1+(1/\gamma))} \\ &\text{(ii)}\quad \|H(u)-\Phi_S^M\|_{L^{\gamma+1}(D_T)} \le K_2\|u_t\|^{1/\gamma}_{L^2(D)}\ T^{(1-(1/\gamma))/(2\gamma+2)}(\Delta t)^{1/\gamma}\ , \end{aligned}$$

hold.

Acknowledgment. This research was carried out under National Science Foundation MCS76-84296 A01. The author thanks M.E. Rose for use of the preprint [18]. The ideas of [18] were used fundamentally in section 5.

References

1. R.A. Adams, Sobolev Spaces, Academic Press, New York, 1975.

2. D.G. Aronson, Regularity properties of flows through porous media, SIAM J. Appl. Math., 17(1969), 461-467.

3. __________, Regularity properties of flow through porous media: A counterexample, Ibid., 19(1970),

4. __________, Regularity properties of flows through porous media: The interface, Archiv Rat. Mech. Anal., 37(1970), 1-10.

5. M.P. Benilan, Existence de solutions fortes pour l'équation des milieux poreux, C.R. Acad. Sc. Paris, 285A(1977), 1029-1031.

6. H. Brezis, On some degenerate nonlinear parabolic equations, in Nonlinear Functional Analysis (F.E. Browder, editor), Part I, pp. 28-38, Amer. Math. Soc., Proc. Symp. in Pure Math., 18, Providence, R.I., 1970.

7. H. Brezis and W. Strauss, Semilinear elliptic equations in L^1, J. Math. Soc. Japan, 25(1973), 565-590.

8. L.A. Caffarelli and A. Friedman, The one-phase Stefan problem and the porous medium diffusion equation: Continuity of the solution in n space dimensions, Proc. Nat. Acad. Sc., 75(1978), 2084.

9. I. Ekeland and R. Temam, Convex Analysis and Variational Problems, North Holland and American Elsevier, Amsterdam and New York, 1976.

10. L.C. Evans, Applications of nonlinear semigroups to partial differential equations, Proceedings of the 1977 Madison Conference on Nonlinear Equations of Evolution, to be published by Academic Press.

11. J.W. Jerome, Nonlinear equations of evolution and a generalized Stefan problem, J. Differential Equations, 26(1977), 240-261.

12. A.S. Kalashnikov, On the occurrence of singularities in the solutions of the equation of nonstationary filtration, Z. Vych. Mat. i. Mat. Fisiki, 7(1967), 440-444.

13. B.F. Knerr, The porous medium equation in one dimension, Trans. Amer. Math. Soc., 234(1977), 381-415.

14. S.N. Kruzkhov, Results on the character of the regularity of solutions of parabolic equations and some of their applications, Math. Z., 6(1969), 97-108.

15. E.B. Leach and M. Sholander, Extended Mean Values, Amer. Math. Monthly, 85(1978), 84-90.

16. J.L. Lions, Quelques Methodes de Resolution des Problemes aux Limites non Lineaires, Dunod, Paris, 1969.

17. O.A. Oleinik, A.S. Kalashnikov and C. Yui-Lin, The Cauchy problem and boundary problems for equations of the type of nonstationary filtration, Izvest. Akad. Nauk SSSR, Ser. Math., 22(1958), 667-704.

18. M.E. Rose, Numerical methods for a porous medium equation, Report ANL-78-80, Argonne National Laboratory, Argonne, Illinois, August, 1978.

19. __________, Numerical methods for a general class of porous medium equations, Argonne National Laboratory Report, Argonne, Illinois, 1979.

20. E.S. Sabinina, On the Cauchy problem for the equation of nonstationary gas filtration in several space variables, Doklady Akad. Nauk SSSR, 136(1961), 1034-1037.

21. G. Stampacchia, Equations Elliptic du Second Ordre a Coefficients Discontinus, University of Montreal Press, Montreal 1966.

22. W.A. Strauss, On weak solutions of semi-linear hyperbolic equations, An. Acad. Brasil. Cienc., 42(1970), 645-651.

NUMERICAL METHODS FOR PHASE-PLANE PROBLEMS IN ORDINARY DIFFERENTIAL EQUATIONS

J.D. Lambert and R.J.Y. McLeod

1. PHASE-PLANE PROBLEMS

Many oscillation problems in applied mathematics reduce to the solution of the autonomous second order scalar equation

$$\ddot{x} + f(x,\dot{x})\dot{x} + g(x) = 0 \; , \quad x = x(t) \; , \quad \dot{x} = \frac{dx}{dt} \; .$$

By writing $\dot{x} = y$ this can be reduced to the first order 2×2 systems

$$\left.\begin{array}{l} \dot{x} = p(x,y) \; , \quad \dot{y} = q(x,y) \\ \text{or} \\ \dfrac{dx}{p(x,y)} = \dfrac{dy}{q(x,y)} \; (= dt) \end{array}\right\} . \tag{1.1}$$

The x-y plane is then called the phase-plane. An elementary example is afforded by the simple harmonic oscillator described by

$$\dot{x} + x = 0 \tag{1.2}$$

for which the corresponding system in the phase-plane is

$$\frac{dx}{x} = - \frac{dy}{y} \; . \tag{1.3}$$

The solutions of (1.2) consist of sinusoidals in the x-t plane, while those of (1.3) consist of circles centre the origin in the x-y plane. Note that it is not always necessary to choose $\dot{x} = y$ to obtain the system (1.1), that is, literally to interpret the phase-plane as being the "displacement-velocity" plane. Thus, for example, the equation $\ddot{x} - 6x\dot{x} + 4x^3 = 0$ is arguably more conveniently represented by the system $\dot{x} = x^2 - y^2$, $\dot{y} = 2xy$ than by the system $\dot{x} = y$, $\dot{y} = 6xy - 4x^3$.

We shall use the phrase "phase-plane problem" to refer to any problem of the form (1.1) where the aim is to find not solutions in the x-t and y-t planes but the trajectories, that is, the integral curves in the x-y plane. The literature contains many analytical results concerning trajectories (see, for example, Davies and James [1]), but for many problems of interest trajectories cannot be found by analytical methods. The object of this paper is to consider the use of numerical methods for finding trajectories, and to propose some new methods which are naturally suited to this problem.

2. SPLIT SCALAR PROBLEMS

The ideas of this paper can also be applied to the problem of finding the integral curve of the scalar initial value problem

$$y' = f(x,y) \ , \quad y(x_0) = x_0 \ , \quad y' = \frac{dy}{dx} \tag{2.1}$$

which can clearly be replaced by the phase-plane problem

$$\dot{x} = p(x,y) \ , \quad \dot{y} = q(x,y) \ , \quad x = x_0 \text{ when } y = y_0 \ , \tag{2.2}$$

where $f(x,y) \equiv q(x,y)/p(x,y)$. We assume that p and q have no common factor; otherwise unnecessary analytical difficulties may be introduced. Thus, for example, the equation $y' = -x/y$ can be split into $\dot{x} = y$, $\dot{y} = -x$, whose trajectories are circles centre the origin; the splitting $\dot{x} = xy$, $\dot{y} = -x^2$, however, introduces a singular line $x = 0$, past which numerical methods fail to integrate.

That advantages can accrue from splitting an initial value problem in this manner is illustrated by the example

$$y' = \frac{x+y}{x-y} \ , \quad y(x_0) = y_0 \tag{2.3}$$

which has solution

$$y = x \tan \ln(C(x^2+y^2)^{1/2}) \ , \quad C = (x_0^2+y_0^2)^{-1/2}\exp(\tan^{-1} y_0/x_0) \ . \tag{2.4}$$

Plotting this integral curve in the x-y plane involves solving by some iterative process a transcendental equation for y at each chosen data point x . A phase-plane problem equivalent to (2.3) is

$$\dot{x} = x - y \ , \quad \dot{y} = x + y \ , \quad x = x_0 \text{ when } y = y_0 \ . \tag{2.5}$$

The general solution of the differential system in (2.5) is

$$x = e^t(A \cos t + B \sin t) \ , \quad y = e^t(-B \cos t + A \sin t) \ . \tag{2.6}$$

We can choose the origin of t at will, and choosing $t = 0$ at $x = x_0$, $y = y_0$ specifies the arbitrary constants to be $A = x_0$, $B = -y_0$. The integral curve in the x-y plane is now readily plotted. Indeed, (2.6) provides a natural parametrization of (2.4) - one that does not suggest itself immediately from the form of (2.4)!

Advantages also accrue when we use numerical methods to solve (2.5) rather than (2.3). Firstly, conventional numerical methods presume that the function f in (2.1) satisfies a Lipschitz condition, and that (2.1) thus uniquely defines y as a <u>function</u> of x , whereas the solution (2.4) merely defines a <u>relation</u> between x and y ; it is this relation which we wish to plot as a trajectory. Indeed the trajectory defined by (2.4) is a spiral, and numerical methods applied to (2.3) fail to integrate past the point where this spiral is cut by the line $y = x$. The system (2.5) suffers no such disadvantage and defines x and y as functions of t everywhere except at the point $x = y = 0$, a singular point of the system through which all trajectories pass. Secondly, (2.5) is linear whereas (2.3) is nonlinear, and calls for substantially more computational effort if an implicit method is used. A third point concerns the distribution of the data points. If, for example, a graph plotter is to be used to present an approximate trajectory, then it is clearly inappropriate to have data evenly spaced in x, as would happen if we applied a method with fixed step to (2.3). It is not clear, a priori, what distribution would be best; one in which the data is equally spaced in terms of arc length along the trajectory is not ideal, since it is desirable to have more points per unit arc length in regions where the curvature is large. As a first attempt, data equally spaced in terms of a parameter, as would be afforded by a method applied with fixed step to (2.5), would appear to be a reasonable strategy.

Clearly, the splitting involved in the replacement of (2.1) by (2.2) is not unique. One of the difficulties in attempting to assess the performance of conventional numerical methods applied to the split form of (2.1) is that the numerical solution is not, in general, invariant with respect to all splittings. One could introduce a "normalized splitting" such as $\dot{x} = p/(p^2+q^2)^{1/2}$, $\dot{y} = q/(p^2+q^2)^{1/2}$, but this substantially increases the computational effort involved; moreover, since it precludes the use of a sensible linear test equation, it stifles attempts to analyse the performance of the numerical method. The methods to be proposed in this paper have the advantage that they are invariant under all splittings of the given problem (2.1).

3. INTERPOLATORY DERIVATION OF NUMERICAL METHODS

The methods to be proposed have properties which are stronly related to the geometry of the underlying local interpolants. In this section we briefly review the connection between local interpolation and a class of conventional numerical methods, the linear multistep methods.

Let $x_n = x_0 + nh$, h the steplength, let the solution of the initial value problem $y' = f(x,y)$, $y(x_0) = y_0$, be locally represented on $[x_n, x_{n+1}]$ by an interpolant $I(x)$, and impose the interpolating conditions

$$I(x_{n+j}) = y_{n+j}, \quad I'(x_{n+j}) = f_{n+j}, \quad j = 0, 1 . \tag{3.1}$$

If we choose $I(x)$ to be a quadratic polynomial,

$$I(x) = ax^2 + bx + c \tag{3.2}$$

then any three of the conditions (3.1) determine the parameters a, b, c, and substitution in the fourth condition yields the method

$$y_{n+1} - y_n = \frac{h}{2}(f_{n+1}+f_n) .$$

Since the method must be exact for any problem whose solution is a quadratic in x, the order is at least two. Thus the Trapezoidal Rule can be interpreted as local Hermite interpolation at two points by a quadratic polynomial in x. (Note that there is no one-to-one correspondence between interpolants and methods; Simpson's

Rule can be interpreted as local Hermite interpolation by a quartic polynomial or a cubic spline interpolant.)

One difficulty in this method of derivation of linear multistep methods is that it yields only methods of maximal order which, by a well-known result, are zero-unstable when the stepnumber exceeds one for an explicit method or two for an implicit. For example, if the interpolant is chosen to be

$$I(x) = ax^3 + bx^2 + cx + d \tag{3.3}$$

the parameters are specified by the four conditions (3.1) and extrapolation is performed by requiring that $y_{n+2} = I(x_{n+2})$, we obtain the third order explicit method

$$y_{n+2} + 4y_{n+1} - 5y_n = h(4f_{n+1}+2f_n)$$

which is zero-unstable. This difficulty can be overcome by replacing the cubic interpolant (3.3) by the quadratic (3.2) and requiring as interpolating conditions

$$\begin{aligned} -(1+\beta)I(x_{n+1}) + \beta I(x_n) &= -(1+\beta)y_{n+1} + \beta y_n \\ I'(x_{n+j}) &= f_{n+j}, \quad j = 0, 1. \end{aligned} \tag{3.4}$$

The interpolant (3.2) is now specified, and extrapolating to x_{n+2} yields the second order method

$$y_{n+2} - (1+\beta)y_{n+1} + \beta y_n = \frac{h}{2}[(3-\beta)f_{n+1} - (1+\beta)f_n] \tag{3.5}$$

in which the parameter β can be chosen to avoid zero instability. Note that if $\beta = 0$ (Adams-Bashforth method) then $I(x)$ interpolates y_{n+1} but not y_n ; if $\beta = -1$ (Mid-point method) then $I(x)$ interpolates y_n but not y_{n+1} ; if $\beta \neq 0, -1$ then $I(x)$ interpolates neither y_n nor y_{n+1} .

By using interpolating conditions similar to (3.4), any linear multistep method may be interpreted in terms of local Hermite interpolation by a polynomial in x , whose degree is less than or equal to the order of the method.

4. METHODS BASED ON A GENERAL PARABOLA

We have seen that the linear multistep method (3.5) is equivalent to locally interpolating by the parabola $y = ax^2 + bx + c$. The axis of this parabola always points in the direction of the y-axis, and it is inevitable that the associated method will run into difficulties when the slope of the integral curve becomes infinite. We propose locally to rotate the interpolating parabola and to retain as parameter γ , the slope of the axis of the parabola relative to the fixed x,y-axes.

Consider the local rotation of axes illustrated in Fig. 1, where $\tan\theta = \gamma := \mu/\nu$.

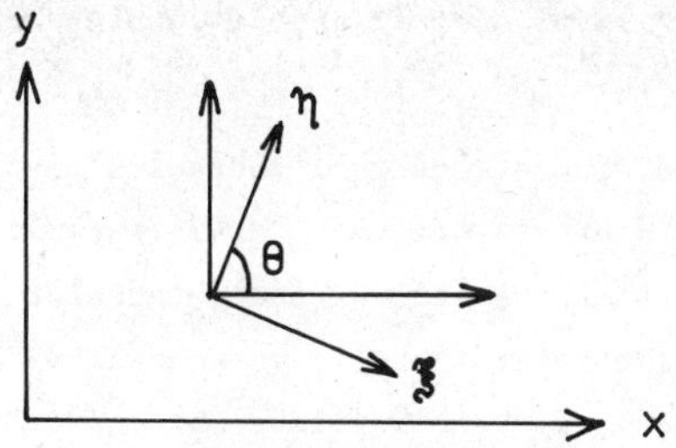

Fig. 1

The transformation $(x,y) \to (\xi,\eta)$ and its inverse are given by

$$\begin{aligned} \xi &= (\mu x - \nu y)/\lambda \qquad & x &= (\mu\xi + \nu\eta)/\lambda \\ \eta &= (\nu x + \mu y)/\lambda \qquad & y &= (\mu\eta - \nu\xi)/\lambda \end{aligned} \quad , \quad \lambda = (\mu^2+\nu^2)^{1/2} \tag{4.1}$$

Consider the phase-plane problem

$$\dot{x} = p(x,y) \ , \quad \dot{y} = q(x,y) \tag{4.2}$$

and the associated differential equation

$$\frac{dy}{dx} = f(x,y) \equiv q(x,y)/p(x,y) \ , \tag{4.3}$$

and let the equation obtained by applying the transformation (4.1) to (4.3) be

$$\frac{d\eta}{d\xi} = \phi(\xi,\eta) \ . \tag{4.4}$$

We now apply the linear multistep method (3.5) to (4.4), but relative to the ξ,η-axes to obtain

$$\begin{aligned} &\eta_{n+2} - (1+\beta)\eta_{n+1} + \beta\eta_n = \frac{h}{2}[(3-\beta)\phi_{n+1} - (1+\beta)\phi_n] \\ &\xi_{n+2} = \xi_{n+1} + h \ , \quad \xi_{n+1} = \xi_n + h \ . \end{aligned} \tag{4.5}$$

Applying the transformation (4.1) to (4.5) and setting

$$\begin{aligned} \phi_{n+j} = \left.\frac{d\eta}{d\xi}\right|_{n+j} &= (\nu+\mu f_{n+j})/(\mu-\nu f_{n+j}) \\ &= (\nu p_{n+j}+\mu q_{n+j})/(\mu p_{n+j}-\nu q_{n+j}) \ , \quad j = 0, 1 \ , \end{aligned}$$

yields the following explicit method for the system (4.2):

$$\left.\begin{aligned} &\lambda^2 x_{n+2} = \nu[(1+\beta)Y_{n+1} - \beta Y_n] + h_n\lambda[\tfrac{1}{2}(3-\beta)\nu F_{n+1} - \tfrac{1}{2}(1+\beta)\nu F_n + \mu] + \mu G_n \\ &\lambda^2 y_{n+2} = \mu[(1+\beta)Y_{n+1} - \beta Y_n] + h_n\lambda[\tfrac{1}{2}(3-\beta)\mu F_{n+1} - \tfrac{1}{2}(1+\beta)\mu F_n - \nu] - \nu G_n \\ &\text{where} \\ &Y_n = \nu x_n + \mu y_n \ , \quad F_n = (\nu p_n+\mu q_n)/(\mu p_n-\nu q_n) \ , \quad G_n = \mu x_{n+1} - \nu y_{n+1} \\ &h_n = (\mu\Delta x_n - \nu\Delta y_n)/\lambda \ , \quad \lambda^2 = \mu^2 + \nu^2 \end{aligned}\right\} \tag{4.6}$$

Since the rotation is to be a local one, μ and ν , which are yet to be determined, will in general depend on n .

Note that (4.6) is homogeneous in p and q , and so is invariant under all splittings of $y' = f(x,y)$ into $\dot{x} = p(x,y)$, $\dot{y} = q(x,y)$. Note also that if we set $\nu = 0$, that is, demand that the axis of the interpolating parabola be everywhere parallel to the y-axis, the first of (4.6) reduces to $x_{n+2} = x_{n+1} + h_n$, and the second reduces to (3.5), with f replaced by q/p .

The local truncation error of (4.6) is

$$\frac{1}{12}(5+\beta)h_n^3\lambda^4[(\mu p-\nu q)(p\ddot{q}-\ddot{p}q) - 3(\mu\dot{p}-\nu\dot{q})(p\dot{q}-\dot{p}q)]/(\mu p-\nu q)^2 + O(h_n^4)$$

where $\dot{p} = \frac{d}{dt}\, p(x,y)\Big|_{x_n,y_n}$ etc.

5. CHOICE OF PARAMETERS

The method (4.6) is homogeneous in μ and ν ; we thus have two parameters at our disposal, namely β and the ratio $\mu:\nu$.

Choice of β

For the linear multistep method (3.5), the parameter β is chosen to satisfy

$$-1 \le \beta < 1 , \qquad (5.1)$$

a choice which guarantees zero-stability of the method. One interpretation of the property of zero-stability is that it is the ability of the method stably to integrate the test problem $y' = 0$, $y(x_0) = y_0$, whose solution is a straight line parallel to the x-axis. Noting that $\dot{x} = 0$, $\dot{y} = 0$ does not define a family of trajectories, we take as an analogous phase-plane test problem

$$\dot{x} = p , \quad \dot{y} = q , \quad p,q \text{ constants}, \quad x = x_0 \text{ when } y = y_0 \qquad (5.2)$$

whose solution is the straight line

$$y - y_0 = k(x-x_0) , \quad k = q/p . \qquad (5.3)$$

Method (4.6) is applied to this test problem with any choice of μ/ν other than $\mu/\nu = k$; that is, we preclude the case where the axis of the interpolating parabola is chosen to be parallel to the solution curve. The following difference equations are obtained, where E is the forward shift operator defined by $Ez_n := z_{n+1}$:

$$F(E)x_n = \nu P(E)y_n , \quad G(E)y_n = -\mu k P(E)x_n , \qquad (5.4)$$

where

$$F(E) = (\mu-\nu k)E^2 - [2\mu - (1+\beta)\nu k]E + (\mu-\beta\nu k)$$

$$G(E) = (\mu-\nu k)E^2 + [2\nu k - (1+\beta)\mu]E + (\beta\mu-\nu k)$$

$$P(E) = (\beta-1)(E-1) .$$

Case $\beta = 1$

The solution of (5.4) satisfying starting values $(x_0,y_0),(x_1,y_1)$ is

$$x_n = x_0 + n\Delta x_0 \ , \quad y_n = y_0 + n\Delta y_0 \ , \tag{5.5}$$

which clearly lies on the solution line (5.3) if the additional starting value (x_1,y_1) does. However, if the starting values are perturbed so that $\Delta y_0 - k\Delta x_0 = \varepsilon$, then (5.5) satisfies

$$y_n - y_0 = k(x_n - x_0) + n\varepsilon \ ,$$

and we condlude that (4.6) does not integrate (5.2) stably.

Case $\beta \neq 1$

On eliminating either y_n or x_n from (5.4) we obtain the difference equation $\mathcal{L}(E)z_n = 0$ (where $z_n = x_n$ or y_n) whose characteristic polynomial turns out to be

$$\mathcal{L}(r) = (\mu^2 - \nu^2 k^2)(r-1)^3(r-\beta) \ .$$

The corresponding general solutions for x_n and y_n are

$$\begin{aligned} x_n &= A + Bn + Cn^2 + D\beta^n \\ y_n &= a + bn + cn^2 + d\beta^n \ , \end{aligned} \tag{5.6}$$

where A, B, ..., d are arbitrary constants. The presence of a triple root of the characteristic polynomial at $r = 1$ does not, however, imply instability, for on substituting (5.6) into (5.4), we find that the general solution of the latter is

$$\begin{aligned} x_n &= A + Bn + \nu K\beta^n \\ y_n &= a + kBn + \mu K\beta^n \end{aligned} \tag{5.7}$$

The four arbitrary constants in (5.7) are specified in terms of the starting values $(x_0,y_0),(x_1,y_1)$ as follows:

$$K = \frac{\Delta y_0 - k\Delta x_0}{(\beta-1)(\mu-\nu k)} \ , \quad B = \frac{\mu\Delta x_0 - \nu\Delta y_0}{\mu-\nu k} \ , \ A = x_0 - \nu K \ , \ a = y_0 - \mu K \ .$$

If the starting values lie on the solution line (5.3), then (5.7) reduces to (5.5), and the point (x_n,y_n) lies on (5.3). If the starting values are perturbed so that $\Delta y_0 - k\Delta x_0 = \varepsilon$, then (x_n,y_n) satisfies

$$y_n - y_0 = k(x_n - x_0) + \frac{\beta^n - 1}{\beta - 1}\,\varepsilon \ ,$$

and we conclude that (4.6) integrates (5.2) stably if $|\beta| \le 1$. Combining both cases, we conclude that an appropriate range for β is $-1 \le \beta < 1$, which coincides exactly with the corresponding condition (5.1) for the linear multistep method (3.5).

A particular choice for β which yields advantages to be discussed later is $\beta = -1$; (4.6) then becomes a local rotation of a symmetric method, namely the mid-point rule. We note in passing that symmetric linear multistep methods for second order equations have been shown to be advantageous for orbit problems (Lambert and Watson [2]).

Choice of $\mu:\nu$

The method (4.6) fails if the denominator in the expression for F_n or F_{n+1} vanishes. Choosing $\mu:\nu = -p_{n+1}:q_{n+1}$ by setting

$$\mu = p_{n+1} \; , \; \nu = -q_{n+1} \tag{5.8}$$

ensures that the denominator in F_{n+1} never vanishes except at $p = q = 0$, a singular point of the system (4.2) through which we do not attempt to integrate; the denominator in F_n will not vanish for sufficiently small step h_n . Note that in the case $\beta = -1$, the term in F_n vanishes.

The choice (5.8) has, however, more fundamental implications; it means that the axis of the interpolating parabola is always normal to the solution curve through the point (x_{n+1}, y_{n+1}) , a geometrical property we shall make use of later.

With the choice (5.8), the method (4.6) becomes

$$\begin{aligned}(p_{n+1}^2+q_{n+1}^2)x_{n+2} &= q_{n+1}A_n - H_n(q_{n+1}B_n-p_{n+1}) + p_{n+1}M_{n+1}\\ (p_{n+1}^2+q_{n+1}^2)y_{n+2} &= -p_{n+1}A_n + H_n(p_{n+1}B_n+q_{n+1}) + q_{n+1}M_{n+1}\end{aligned} \tag{5.9}$$

where

$$\begin{aligned}A_n &= -\beta(q_{n+1}x_n-p_{n+1}y_n) + (1+\beta)(q_{n+1}x_{n+1}-p_{n+1}y_{n+1})\\ B_n &= \tfrac{1}{2}(1+\beta)(q_{n+1}p_n-p_{n+1}q_n)/(p_{n+1}p_n+q_{n+1}q_n)\\ H_n &= p_{n+1}\Delta x_n + q_{n+1}\Delta y_n \; , \quad M_{n+1} = p_{n+1}x_{n+1} + q_{n+1}y_{n+1} \; .\end{aligned}$$

An alternative form of (5.9), more convenient for computation, is

$$(p_{n+1}^2+q_{n+1}^2)\Delta^2 x_n = q_{n+1}L_{n+1} \; , \; (p_{n+1}^2+q_{n+1}^2)\Delta^2 y_n = -p_{n+1}L_{n+1} \tag{5.10}$$

where

$$L_{n+1} = (1-\beta)(p_{n+1}\Delta y_n-q_{n+1}\Delta x_n) - H_n B_n \; ,$$

with H_n, B_n as given in (5.9).

In the case $\beta = -1$, (5.10) reduces further to

$$\begin{aligned}(p_{n+1}^2+q_{n+1}^2)\Delta x_{n+1} &= (p_{n+1}^2-q_{n+1}^2)\Delta x_n + 2p_{n+1}q_{n+1}\Delta y_n\\ (p_{n+1}^2+q_{n+1}^2)\Delta y_{n+1} &= 2p_{n+1}q_{n+1}\Delta x_n - (p_{n+1}^2-q_{n+1}^2)\Delta y_n \; .\end{aligned} \tag{5.11}$$

Note that the steplength

$$h_n = (p_{n+1}\, x_n + q_{n+1}\, y_n)/(p_{n+1}^2 + q_{n+1}^2)^{1/2} \qquad (5.12)$$

does not always appear explicitly. The above formulae (5.9), (5.10) and (5.11) imply that the steplength is held constant in the following sense; h_n is the perpendicular distance from either (x_n,y_n) or (x_{n+2},y_{n+2}) to the line through (x_{n+1},y_{n+1}) normal to the solution curve through (x_{n+1},y_{n+1}). Note also that once the starting values (x_0,y_0) and (x_1,y_1) are specified, we have no further control over the steplength. However, a geometric interpretation of h_n (see Section 6) shows that as the curvature of the solution curve increases, the data points become more closely packed. In regions of low curvature, the steplength can be doubled by replacing (x_n,y_n) by (x_{n-1},y_{n-1}). However, in this paper we are concerned only with fixed-step applications.

6. EXACTNESS AND STABILITY

For linear multistep methods zero-stability is a necessary but not sufficient condition for the method to perform satisfactorily. In theories of numerical stability for such methods we examine their performances on a test equation, usually taken to be $y' = \lambda y$, λ constant. It is important to note that the motivation for this choice of test equation is not that the exponential function is in any sense an acceptable local model for the solution of a general problem, but that the equation $y' = \lambda y$ is a model for a local linearization of $y' = f(x,y)$, in which we hold $\partial f/\partial y$ constant.

For phase-plane problems the situation is somewhat different. Local linearization of $\dot{x} = p(x,y)$, $\dot{y} = q(x,y)$ leads to several different cases, and, in any event, we are primarily interested in the ability of the method to integrate a trajectory for which y is not a function of x (or vice versa). It seems more appropriate to use a test equation whose solution is a reasonable local model for such a trajectory; in other words, we think in terms of a test solution rather than a test equation. We choose, in the first instance, a circular trajectory as an appropriate test solution. (The use of a circular trajectory as a test solution has already been proposed by Stiefel and Bettis [3] in connection with the numerical calculation of orbits.)

A difficulty with this approach is that many different phase-plane problems have families of circles as trajectories. Consider, for example, the system

$$\dot{x} = y\ ,\quad \dot{y} = -x\ , \qquad (6.1)$$

whose trajectories are a family of concentric circles, and the problem

$$\dot{x} = x^2 - y^2 + 1\ ,\quad \dot{y} = 2xy\ , \qquad (6.2)$$

whose trajectories are a family of non-intersecting coaxal circles. The Trapezoidal Rule has the property that when it is applied to (6.1) it yields data points which lie exactly on the circular trajectory, but this is not the case when it is applied to (6.2). However, method (5.11) (which, note, is explicit whereas the Trapezoidal

Rule is implicit) gives data points which lie exactly on the circular trajectory for either (6.1) or (6.2), provided that the necessary additional starting value also lies on the trajectory. It is of interest to compare the numerical results of these two methods when applied to (6.2) with initial condition $x_0 = \sqrt{2}$, $y_0 = 1$. In Fig. 2 the crosses denote the data points given by the Trapezoidal Rule with fixed step of 0.2 and the circles denote the data points given by (5.11) with additional starting value $x_1 = \sqrt{2.64}$, $y_1 = 1.2$. This, by (5.12), corresponds to $h_0 = 0.448$ (= h_n, for this problem, for all n).

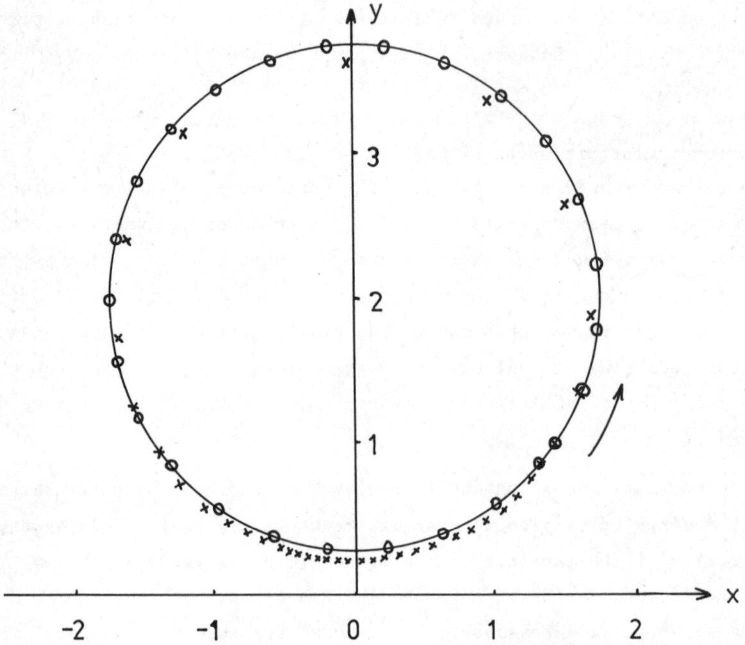

Fig. 2

The exact trajectory is $x^2 + (y-2)^2 = 3$. The data points given by (5.11) lie exactly on this circle and, moreover, are regularly spaced round it. Those given by the Trapezoidal Rule fail to lie on the circle, and are very irregularly spaced. If the Trapezoidal Rule were to be used to find an approximate trajectory with reasonably evenly spaced points, then much step-changing would need to be done.

In view of the above performance of method (5.11), one suspects that it may be exact on any circular trajectory, and we now show this to be the case.

Definition

A numerical method is said to be circularly exact if when applied with exact starting values to any phase-plane problem whose solution trajectory is a circle it yields data points lying exactly on the circular trajectory.

Theorem

Method (5.11) is circularly exact.

Proof

Method (5.11) is method (4.6) with $\beta = -1$, $\mu = p_{n+1}$, $\nu = -q_{n+1}$. Let A,B,C be the points $(x_n,y_n),(x_{n+1},y_{n+1}),(x_{n+2},y_{n+2})$ respectively, and let BB' be a line through B parallel to the axis of the interpolating parabola.

If $\beta = -1$, the interpolating conditions associated with (4.6) (see (3.4)) imply that

(i) the parabola passes through A and C but not necessarily through B ;

(ii) if BB' meets the parabola in P , then the slope of the tangent to the parabola at P equals the slope of the local integral curve at B ;

(iii) the perpendicular distances of A and C from BB' are equal.

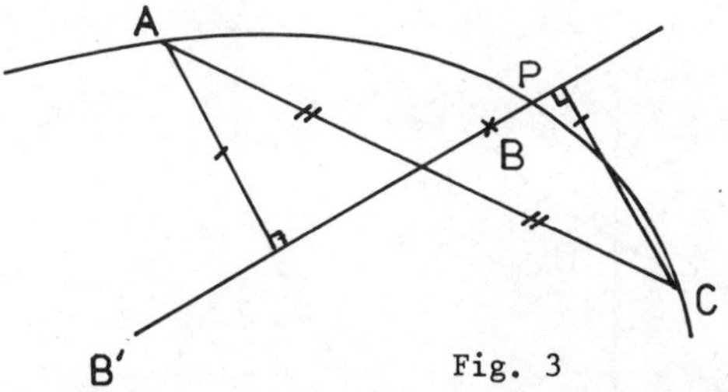

Fig. 3

From Fig. 3 it follows by congruent triangles that BB' bisects AC and, by an elementary property of the parabola, the tangent at P is parallel to AC. Hence, the slope of the integral curve through B is that of AC.

We now impose the conditions $\mu = p_{n+1}$, $\nu = -q_{n+1}$, which imples that BB' is normal to the solution curve through B and hence is perpendicular to AC ; P must then be the vertex of the parabola and we have the situation of Fig. 4.

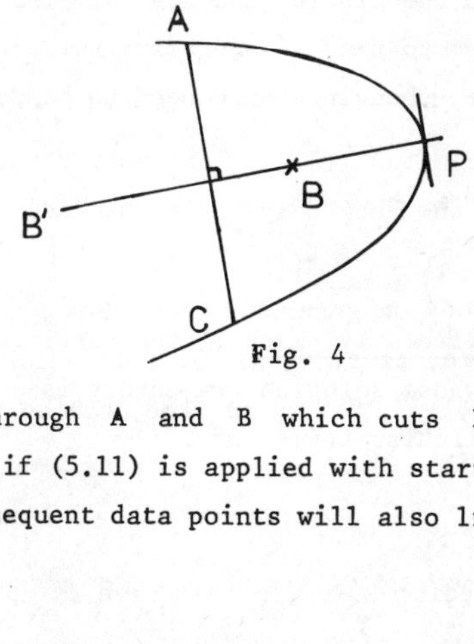

Fig. 4

Clearly the circle through A and B which cuts BB' normally must pass through C . It follows that if (5.11) is applied with starting values which lie on a fixed circle, then all subsequent data points will also lie on the circle. □

We have shown that (5.11) will exactly reproduce any circular trajectory. However, as in the discussion in the first part of Section 3, we have to consider whether this integration round a circular trajectory is numerically stable. If round-off error is present, is it propagated in a stable manner? Numerical experiments in which error is artificially introduced indicate that the process is numerically stable, and this observation is confirmed by the following geometric argument.

We make the assumption that in a small neighbourhood of any point on the given circular trajectory, neighbouring trajectories can be represented by circles concentric with the given circle. We first consider the propagation of a single round-off error. Let C denote the given circular trajectory, and let A_n be the point (x_n, y_n) for all n. Suppose that A_n and A_{n+1} lie on C, but, due to round-off error, A_{n+2} is perturbed to lie on a neighbouring trajectory C'. Let the radius of C through A_{n+2} meet C in B_{n+2} (see Fig. 5).

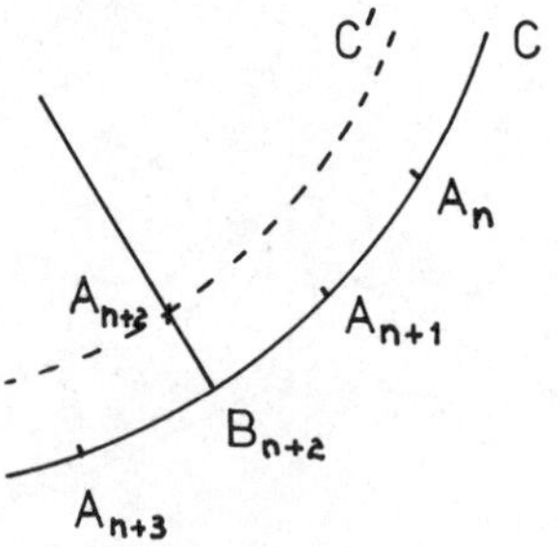

Fig. 5

At the next integration step, the interpolating parabola through A_{n+1} does not pass through A_{n+2} but its slope at its vertex is that of C' at A_{n+2}, which is equal to the slope of C at B_{n+2}. Hence, by the circular exactness of the method, the point A_{n+3} lies on C. A similar argument shows that A_{n+4} lies on C' and subsequent points lie alternately on C and C'. Ths single round-off error at the $(n+2)$ stage thus cannot be amplified. Note that tangential components of round-off error do not affect the radial stability, but they will affect the distribution of points on the trajectory.

In the case when round-off error is introduced at every step, an obvious extension of the above argument shows that if the radial component of the round-off error at each step is bounded by ε, then the radial component of round-off error after N steps is bounded by $N\varepsilon$.

Numerical experiments so far performed on general phase-plane problems corroborate the conclusions of this and the preceding section. In particular, when method (5.10), which assumes the choice $\mu = p_{n+1}$, $\nu = -q_{n+1}$, is used, the best results do appear to be obtained with the choice $\beta = -1$.

7. A NUMERICAL EXAMPLE

Method (5.11) and the Ordinary Mid-point Rule are tested on the problem

$$\dot{x} = y(2x^2+y^2) \ , \ \dot{y} = -x^3 \ , \ x = 0 \ \text{ at } \ y = 1 \ .$$

The exact solution is

$$x^2 + y^2 = \exp(x^2/(x^2+y^2)) \ .$$

For both methods, the additional starting value is obtained by applying an explicit second-order Runge-Kutta method. In Fig. 6, circles denote the points given by method (5.11) and crosses those given by the Mid-point Rule.

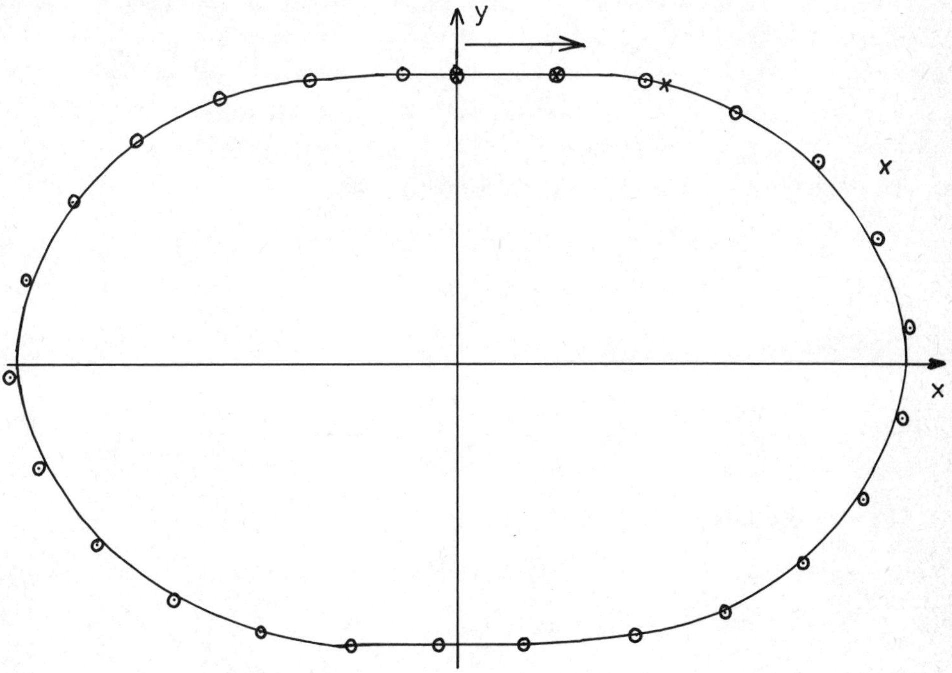

Fig. 6

Note that in the neighbourhood of the initial value the trajectory is almost straight and parallel to the x-axis, so that initially the two methods are almost identical. However, the superior properties of (5.11) are well demonstrated; although not exact, it gives a good representation of the trajectory, with evenly spaced data points and no adverse accumulation of error. The Mid-point Rule, on the other hand, soon becomes unstable, the points following those shown in Fig. 5 being approximately (3,-1), (-11,-17).

8. ON THE EXISTENCE OF A CONICALLY EXACT METHOD

The choice $\mu = p_{n+1}$, $\nu = -q_{n+1}$, which forces the axis of the interpolating parabola to be normal to the local integral curve at (x_{n+1}, y_{n+1}) , resulted in (4.6), with $\beta = -1$, being circularly exact. In this section, we briefly indicate another advantageous choice which can be made for the ratio $\mu:\nu$.

Recall that the situation depicted by Fig. 3 holds for general μ/ν . Using the notation defined in connection with Fig. 3, suppose that A and B lie on a fixed conic, and that we are able to choose $\mu:\nu$ such that BB' passes through the centre of this conic. Since the radius of a conic at a point B on it bisects chords parallel to the tangent to the conic at B , it follows that C also lies on the fixed conic.

Thus it is possible to construct a conically exact method, provided that we are able to choose $\mu:\nu$ at each step such that BB' passes through the centre of the conical trajectory. A conic is determined by five linearly independent conditions, and the points A and B , and the slopes of the integral curve at A and B , are not enough to specify the conical trajectory. Let us assume, in addition, that one back-point, (x_{n-1},y_{n-1}) , also lies on the conic. The most general conic through A and B , with slopes q_n/p_n at A and q_{n+1}/p_{n+1} at B is

$$[p_n(y-y_n) - q_n(x-x_n)][p_{n+1}(y-y_{n+1}) - q_{n+1}(x-x_{n+1})] = \theta[\Delta x_n(y-y_n) - \Delta y_n(x-x_n)]^2. \tag{8.1}$$

This conic passes through (x_{n-1},y_{n-1}) if

$$\theta = \frac{(p_n\Delta y_{n-1}-q_n\Delta x_{n-1})[p_{n+1}(y_{n+1}-y_{n-1}) - q_{n+1}(x_{n+1}-x_{n-1})]}{(\Delta x_n\Delta y_{n-1}-\Delta y_n\Delta x_{n-1})^2} . \tag{8.2}$$

If (8.1) with (8.2) is written in the form

$$ax^2 + 2hxy + by^2 + 2gx + 2fy + c = 0 ,$$

then the centre of the conic has coordinates

$$\frac{hf - bg}{ab - h^2}, \frac{hg - af}{ab - h^2},$$

and the choice

$$\mu = y_{n+1}(ab-h^2) - hg + af$$

$$\nu = x_{n+1}(ab-h^2) - hf + bg$$

ensures that BB' passes through the centre of the conic. The method (4.6) with $\beta = -1$ and this choice for μ and ν will be exact on a conic, provided that the starting values are exact. The argument of the previous section showing numerical stability also applies to conically exact methods, if we make the assumption that in the neighbourhood of a point on the conic, neighbouring trajectories are similar concentric conics.

Acknowledgements

The authors wish to acknowledge the support of NATO, under grant number 1036. They also wish to thank Dr H G Anderson and Professor B D Sleeman of the University of Dundee for helpful comments.

References

[1] Davies, T.V. and James, E.M., "Nonlinear Differential Equations", Addison-Wesley, 1966.

[2] Lambert, J.D. and Watson, I.A., "Symmetric multistep methods for periodic initial value problems", Journal I.M.A., 18, 189-202, 1976.

[3] Steifel, E. and Bettis, D.G., "Stabilization of Cowell's method", Numer. Math., 13, 1969.

ON THE USE OF EXACT PENALTY FUNCTIONS TO DETERMINE STEP LENGTH IN OPTIMIZATION ALGORITHMS

David Q. Mayne

1. INTRODUCTION

Exact penalty functions have been employed in two distinct ways in the mathematical programming literature. In the first (Conn (1973), Conn and Pietrzykowski (1977)) a constrained optimization problem P is replaced by the unconstrained problem P_c of minimizing an exact penalty function $\gamma(x,c)$, where $c \in R$ denotes the penalty parameter. If c is sufficiently large (but finite) a local solution of P_c is also a solution of P. In this class of algorithms the search direction is chosen to be a descent direction for the exact penalty function; the non-differentiability of this function must be taken into account.

More recently (Han (1977), Powell (1978), Mayne and Maratos (1979), Mayne and Polak (1978)) the exact penalty function has been employed to enforce global convergence in algorithms which determine the search direction by solving a first or second order approximation to P. An example is the extension, due to Levitin and Polyak (1966), of Newton's method for constrained optimization problems; this algorithm is only locally convergent but could be made globally convergent by employing an exact penalty function to determine step length. Search directions chosen by solving local approximations to P are not necessarily descent directions for P_c; this fact controls the choice of the penalty parameter c.

This class of algorithms is appealing precisely because of the manner in which the search direction is obtained. In particular, if a second order approximation to P is employed, a superlinear or quadratic rate of convergence may be obtained. However, several difficulties, which have so far received very little attention, need to be faced. The first of these concerns the choice of the penalty parameter c. Han (1977) assumes that a suitable value is known a priori; Powell (1978) gives a rule which appears to work well in most problems. It would appear important, however, to determine rules for choosing c which guarantee global convergence. Secondly, attention must be given to the calculation of step length since the exact penalty function is non-differentiable. We show that an Armijo type procedure has the necessary properties required for convergence. A more subtle difficulty concerns the rate of convergence of the algorithms utilizing second order approximations to P to determine search length. Unless the step length is asymptotically unity, superlinear convergence is not achieved. Han (1977) therefore suggests that the exact penalty step length procedure be discarded near solution points but there is no known automatic procedure for doing so. Maratos (1978) has shown that there exist problems for which the exact penalty procedure produces step lengths of less than unity, no matter how close the current iterate is to a solution point.

This paper, based on results obtained in Mayne and Maratos (1979), Mayne and Polak (1978), and Maratos (1978) shows how these difficulties may be overcome. Two algorithms, a first order algorithm for equality constrained optimization problems and a second order algorithm for constrained optimization, illustrate the proposed solutions.

2. THE OPTIMIZATION ALGORITHMS

The first algorithm solves:

$$P^1 : \min \{f(x) \mid h(x) = 0\}$$

where $f : R^n \to R$ and $h : R^n \to R^r$ are continuously differentiable. It is assumed that the constraint gradients $\nabla h^1(x), \ldots, \nabla h^r(x)$ are linearly independent for all x in R^n. The algorithm determines a search direction by solving a first order approximation to P^1. Thus at x the algorithm generates a search direction p(x) as follows:

$$p(x) = p_1(x) - \omega(x)p_2(x)$$

where $p_1(x)$ is the minimum (Euclidean) norm solution of:

$$\hat{h}(x,p) = 0$$

where

$$\hat{h}(x,p) \triangleq h(x) + h_x(x)p$$

and $p_2(x)$ is the orthogonal projection of the cost gradient $\nabla f(x)$ onto the subspace orthogonal to the subspace spanned by the constraint gradients $\nabla h^1(x), \ldots, \nabla h^r(x)$. The normalising factor $\omega(x)$ is the step length, along the direction $p_2(x)$, which approximately minimizes the Lagrangian $L(x-\omega p_2(x), \lambda(x))$ using a single step quadratic interpolation procedure. The Lagrangian is defined by:

$$L(x,\lambda) \triangleq f(x) - \langle \lambda, h(x) \rangle$$

and the multiplier estimator $\lambda : R^n \to R$ by

$$\lambda(x) = (h_x(x)^T)^+ \nabla f(x)$$

where $(\cdot)^+$ denotes pseudo inverse.

To determine step length a penalty function is required. Among many possibilities (Charalambous (1978)) we choose:

$$\gamma(x,c) \triangleq f(x) + c||h(x)||_1$$

so that the penalty parameter c is a scalar. Our analysis is easily extended to penalty functions like $f(x) + [\Sigma(c^j h^j(x))^p]^{1/p}$ where c is a vector. The step length $\alpha(x)$ is then calculated to minimize $\gamma(x+\alpha p(x), c)$ approximately; the step length clearly depends on the penalty parameter c (and may be zero if c is too small).

The procedures for search direction and step length together define a function $A_c : R^n \to R^n$:

$A_c(x) \triangleq x + \alpha(x)p(x)$

i.e. if x is the current point and c is the current value of the penalty parameter, $A_c(x)$ is the next point generated by algorithm 1. A procedure for choosing c completes the algorithm description.

The second algorithm solves:

$P^2 : \min \{f(x) | g(x) \leq 0, h(x) = 0\}$

where $f : R^n \to R$, $g : R^n \to R^m$ and $h : R^n \to R^r$ are three times continuously differentiable. In order to specify the linear independence condition we define the active constraint sets as follows:

$I(x) \triangleq \{j \in \underline{m} | g^j(x) = \psi(x)\}$

$I_e(x) \triangleq \{j \in \underline{r} | |h^j(x)| = \psi(x)\}$

where $\underline{m}$ denotes the set $\{1,\dots,m\}$ and $\psi : R^n \to R$ is defined by:

$\psi(x) \triangleq \max \{g^j(x), j \in \underline{m}; |h^j(x)|, j \in \underline{r}; 0\}$

The constraint gradients $\{\nabla g^j(x), j \in I(x); \nabla h^j(x), j \in I_e(x)\}$ are assumed to be linearly independent for all x in R^n. The search direction p(x) at x is obtained as the solution to the quadratic program QP(x,H) defined by

$QP(x,H) : \min \{f_x(x)p + \tfrac{1}{2}p^T Hp | g(x) + g_x(x)p \leq 0, h(x) + h_x(x)p = 0\}$.

If the solution is not unique, the minimum norm Kuhn-Tucker triple $\{p,\lambda,\mu\}$ for QP(x,H) is selected, and the search direction p(x) set equal to p. The matrix H is an approximation to L_{xx} where the Lagrangian L is defined by:

$L(x,\lambda,\mu) \triangleq f(x) + \langle\lambda, g(x)\rangle + \langle\mu, h(x)\rangle$.

However a solution to QP(x,H) may not exist or may be unsatisfactory, in which case it may be replaced by a descent direction for the exact penalty function $\gamma(x,c)$; a suitable penalty function is:

$\gamma(x,c) \triangleq f(x) + c\psi(x)$

The step length $\alpha(x)$ is again chosen to minimize approximately the exact penalty function. Since the step length depends on the penalty parameter c, the next iterate is again $A_c(x) = x + \alpha(x)p(x)$. Before continuing the matrix H is updated.

3. ALGORITHM MODEL

In order to design algorithms of this type, it is desirable to obtain properties which, if possessed by the map A_c and the rule for choosing c, guarantee convergence.

It would seem necessary to require that the search direction p(x) generated by the algorithm is a descent direction for the exact penalty $\gamma(x,c)$.

Let $\hat{\gamma}(x,p,c)$ denote the first order, convex, estimate of $\gamma(x+p,c)$ obtained by replacing $f(x+p)$ and $g(x+p)$ in the definition of $\gamma(x+p,c)$ by $\hat{f}(x,p) \triangleq f(x) + f_x(x)p$ and $\hat{g}(x,p) \triangleq g(x) + g_x(x)p$. Then $\theta(x,p,c) \triangleq \hat{\gamma}(x,p,c)-\gamma(x,c)$ is an estimate of $\gamma(x+p,c) - \gamma(x,c)$; p is a descent direction for $\gamma(x,c)$ if $\theta(x,p,c) < 0$. For the first algorithm, since $\hat{h}(x,p(x)) = 0$,

$$\theta(x,p(x),c) = \langle\nabla f(x),p(x)\rangle - c\|h(x)\|_1$$
$$= -\sum_{j=1}^{r} [\lambda^j(x)h^j(x)+c|h^j(x)|] - \omega(x)\|p_2(x)\|^2$$

and a sufficient condition for the search direction $p(x)$ to be a descent direction for $\gamma(x,c)$ is $c \geq \bar{c}(x)$ where:

$$\bar{c}(x) \triangleq b + \max\{|\lambda^j(x)|\ |j \in \underline{r}\}$$

where $b > 0$. For the second algorithm, if $\{p,\lambda,\mu\}$ is a Kuhn-Tucker triple for $QP(x,H)$ and $p(x) = p$, it can be shown that:

$$\theta(x,p(x),c) = \langle\nabla f(x),p(x)\rangle - c\psi(x) \leq -p^T Hp-[c-\sum_{j=1}^{m}\lambda^j-\sum_{j=1}^{r}|\mu^j|]\psi(x)$$

so that $p(x)$ is a descent direction if $H > 0$ and $c \geq \bar{c}(x)$ defined by:

$$\bar{c}(x) \triangleq b + \sum_{j=1}^{m}\lambda^j + \sum_{j=1}^{r}|\mu^j|$$

Let D denote the set of desirable points for the optimization problem. For P^1, D is the set $\{x|h(x) = 0,\ p_2(x) = 0\}$; for P^2, D is the set of Kuhn-Tucker points. If the penalty parameter c remains constant, each algorithm generates a sequence $\{x_i\}$ satisfying $x_{i+1} = A_c(x_i)$ for all i; any accumulation point x^* of such a sequence satisfies $\theta(x^*,p(x^*),c) = 0$. Let D_c be the set of such points. The following algorithm model, which has the same structure as our two algorithms, defines a rule for choosing the penalty parameter; the associated convergence theorem specifies properties for this rule and the map A_c which guarantee convergence.

Algorithm Model

Data: $x_1 \in R^n$, $c_0 > 0$.

Step 0: Set $i = 1$.

Step 1: If $c_{i-1} \geq \bar{c}(x_i)$, set $c_i = c_{i-1}$.

If $c_{i-1} < \bar{c}(x_i)$, set $c_i = \max\{c_{i-1}+\delta, \bar{c}(x_i)\}$.

Step 2: Compute $x_{i+1} = A_{c_i}(x_i)$. Set $i = i+1$.

Go to Step 1. □

Theorem 1

Suppose $\bar{c}$ has the following properties:

(i) $\bar{c} : R^n \to R$ is continuous.

(ii) $x^* \in D_c$ and $c \geq \bar{c}(x^*) \Rightarrow x^* \in D$.

Suppose A_c is such that:

(iii) Any accumulation point x^* of any infinite sequence $\{x_i\}$, for which $x_{i+1} = A_c(x_i)$ and $c \geq \bar{c}(x_i)$ for all i, satisfies $x^* \in D_c$.

Then any sequence $\{x_i\}$ generated by the algorithm model has the following properties:

(a) If the penalty parameter is increased only finitely often, then any accumulation point x^* of $\{x_i\}$ lies in D.

(b) If the penalty parameter c_{i-1} is increased infinitely often when $i \in K = \{i_1, i_2, \ldots\}$, then the subsequence $\{x_i\}_{i \in K}$ has no accumulation points. □

To appreciate the significance of (b) we note the following consequence of Theorem 1.

Corollary

If the sequence $\{x_i\}$ generated by the algorithm model is bounded, then the penalty parameter is increased only finitely often and any accumulation point x^* of $\{x_i\}$ lies in D. □

Proof of Theorem 1

(a) Since the penalty parameter is increased finitely often, the sequence $\{x_i\}$ generated by the algorithm satisfies $x_{i+1} = A_c(x_i)$ for all $i \geq j$ and some finite c. From hypothesis (iii), any accumulation point x^* of $\{x_i\}$ satisfies $x^* \in D_c$. From Step 1 of the algorithm, $c_i = c \geq \bar{c}(x_i)$ for all $i \geq j$; since $\bar{c}$ is continuous, $c \geq \bar{c}(x^*)$. Hence, from hypothesis (ii), $x^* \in D$.

(b) Suppose $x_i \to x^*$ as $i \to \infty$, $i \in \bar{K} \subset K$, so that $c_i \to \infty$. Let ε^* be any positive number and let c^* denote the maximum of $\bar{c}(x)$ over the (closed) ε^*-neighbourhood of x^*. Then, since $c_i \to \infty$ we have $c_{i-1} \geq c^* \geq \bar{c}(x_i)$ for all i sufficiently large, $i \in K$. But $i \in K$ implies that the penalty parameter is increased, i.e. that $c_{i-1} < \bar{c}(x_i)$, a contradiction. Hence $\{x_i\}_{i \in K}$ cannot possess an accumulation point. □

The reasons for the specified properties for $\bar{c}$ and A_c are seen in the proof. Hypothesis (ii) is an obvious requirement.

4. THE TEST FUNCTION $\bar{c}$

For the first algorithm $\bar{c}$ is defined by $\bar{c}(x) = b + \max\{|\lambda^j(x)| \mid j \in \underline{r}\}$, and the continuity of $\bar{c}$ follows from the continuity of $\lambda^j(x)$. If $c \geq \bar{c}(x)$ then, from the definitions of §3:

$$\Theta(x,p(x),c) \leq -b||h(x)||_1 - \omega(x)||p_2(x)||^2$$

If $p_2(x) \neq 0 \to \omega(x) > 0$ (an obvious requirement) then $\Theta(x,p(x),c) = 0$ implies that $h(x) = 0$, $p_2(x) = 0$, i.e. $x \in D_c \Rightarrow x \in D$ so that $\bar{c}$ has the necessary properties.

For the second algorithm, $\bar{c}(x) = b + \Sigma\lambda^j + \Sigma|\mu^j|$, where $\{p,\lambda,\mu\}$ is a Kuhn-Tucker triple for QP(x,H); λ, μ do *not* vary continuously with x. Hence Mayne and Polak (1978) employ $\bar{c}$ defined by:

$$\bar{c}(x) = b + \sum_{j=1}^{m} \bar{\lambda}^j(x) + \sum_{j=1}^{r} |\bar{\mu}^j(x)|$$

where $\bar{\lambda}$ and $\bar{\mu}$ are alternative, continuous estimates of the multipliers with the property that $\bar{\lambda}(\hat{x}) = \hat{\lambda}$, $\bar{\mu}(x) = \hat{\mu}$ if $(\hat{x},\hat{\lambda},\hat{\mu})$ is a Kuhn-Tucker triple for P_2. We also establish that, if $c \geq c(x^*)$, then $x^* \in D_c \Leftrightarrow x^* \in D$. However satisfaction of the test $c \geq \bar{c}(x^*)$ does not now necessarily ensure that the Kuhn-Tucker point p of QP(x,H) is a descent direction for $\gamma(x,c)$.

5. THE MAP A_c

The basic requirement for the map A_c, which generates the next iterate, is hypothesis (iii)of Theorem 1, i.e. any accumulation point of an infinite sequence generated by A_c lies in $D_c = \{x|\Theta(x,p(x),c) = 0\}$. To obtain this property we have to specify a suitable step length procedure. We recall that $\Theta(x,p(x),c)$ is an estimate of $\gamma(x+p(x),c) - \gamma(x)$, and that p(x) is the search direction generated by the algorithm.

Consider the algorithm for P1. It is shown in §4, if $c \geq \bar{c}(x)$, that:

$$\Theta(x,p(x),c) \leq -b||h(x)||_1 - \omega(x)||p_2(x)||^2$$

The definition of ω implies that $\omega(x') \geq a > 0$ for all x' in some neighbourhood of x if $p_2(x) \neq 0$. It follows from the convexity of $\alpha \mapsto \Theta(x,\alpha p(x),c)$, that $\Theta(x,\alpha p(x),c)$, a first order estimate of $\gamma(x+\alpha p(x),c) - \gamma(x,c)$, satisfies:

$$\Theta(x,\alpha p(x),c) \leq \alpha\Theta(x,p(x),c) \quad \text{for all } \alpha \in [0,1].$$

Since the estimation error is $o(\alpha)$ it can be shown (Mayne and Maratos (1979)) that for all $x \notin D_c$ there exists an $\bar{\varepsilon} > 0$ and an $\bar{\alpha} > 0$ such that:

$$\gamma(x'+\alpha p(x'),c)-\gamma(x',c) \leq \alpha\Theta(x',p(x'),c)/2$$

for all $x' \in N_{\bar{\varepsilon}}(x) \triangleq \{x' | \; ||x'-x|| \leq \bar{\varepsilon}\}$, all $\alpha \in [0,\bar{\alpha}]$.

If at x the step length $\alpha(x)$ is chosen to be the largest α in the set $\{1,\beta,\beta^2,\ldots\}$, $\beta \in (0,1)$, such that the inequality $\gamma(x+\alpha p(x),c) - \gamma(x,c) \leq \alpha\Theta(x,p(x),c)/2$ is satisfied, then $\alpha(x') \geq \beta\bar{\alpha}$ for all $x' \in N_\varepsilon(x)$. It follows from the upper semi-continuity of $x \mapsto \Theta(x,p(x),c)$ that there exists an $\varepsilon \in (0,\bar{\varepsilon}]$ and a $\delta > 0$ such that:

$$\gamma(A_c(x'),c) - \gamma(x',c) \leq -\delta$$

for all $x' \in N_\varepsilon(x)$. The algorithm produces a uniform reduction in $\gamma(\cdot,c)$ in a suitably small neighbourhood of any x not lying in D_c. It follows from Theorem (1.3.3) in Polak (1971) that any accumulation point x^* of an infinite sequence $\{x_i\}$, where $x_{i+1} = A_c(x_i)$ and $c \geq \bar{c}(x_i)$ for all i, lies in D_c. The above rule for choosing step length is an extension of the usual Armijo rule, the estimate $\alpha\Theta(x,p,c)$ of $\gamma(x+\alpha p,c) - \gamma(x,c)$ replacing the usual estimate $\alpha\langle\nabla f(x),p\rangle$ of $f(x+\alpha p) - f(p)$.

The algorithm for P_2 is necessarily more sophisticated. Suppose that the current iterate is x and the current estimate of L_{xx} is H. Then the search direction p(x) is set equal to p where $\{p,\lambda,\mu\}$ is the (minimum norm) Kuhn-Tucker triple for QP(x,H) provided certain conditions are met. Otherwise p(x) is set equal to $\bar{p}(x,c)$, a (first order) descent direction for $\gamma(x,c)$ such that, for all c, $x \mapsto \bar{p}(x,c)$ and $x \mapsto \Theta(x,\bar{p}(x,c),c)$ are continuous (and $\Theta(x,\bar{p}(x,c),c) < 0$ for all $x \notin D_c$). Such a direction may be computed by solving a simple quadratic program (Mayne and Polak (1978)). To complete the description we need to specify conditions to be met by the Kuhn-Tucker point of QP(x,H) if it is to be accepted as a search direction. Suitable conditions are:

(α) A solution $\{p,\lambda,\mu\}$ of QP(x,H) exists.

(β) $||p|| < L < \infty$

(γ) $\Theta(x,p,c) \leq -T(x)$

where $T : R^n \to R^+$ is a continuous function such that $T(x) = 0$ if and only if $x \in D_c$, $c \geq \bar{c}(x)$, i.e. if and only if $x \in D$. Test (γ) overcomes the difficulty caused by the lack of continuity of the solution of QP(x,H).

As in Algorithm 1:

$$\hat{\gamma}(x,\alpha p(x),c) - \gamma(x,c) = \Theta(x,\alpha p(x),c)$$

$$\leq \alpha\Theta(x,p(x),c) \quad \text{for all } \alpha \in [0,1]$$

where $\Theta(x,p(x),c) \leq -T(x)$ if the tests are satisfied and $\Theta(x,p(x),c) = \Theta(x,\bar{p}(x,c),c)$ otherwise. From the continuity properties of T, $\bar{p}$ and Θ it can be shown, as before, that there exists an $\bar{\varepsilon} > 0$ and an $\bar{\alpha} \in (0,1]$ such that:

$$\gamma(x'+\alpha p(x'),c) - \gamma(x',c) \leq \alpha\Theta(x',p(x'),c)/8$$

for all $x' \in N_{\bar{\varepsilon}}(x)$, all $\alpha \in [0,\bar{\alpha}]$. If the (extended) Armijo step length rule is employed then $\alpha(x) \geq \beta\bar{\alpha} > 0$ for all $x' \in N_{\bar{\varepsilon}}(x)$; also, there exists an $\varepsilon \in (0,\bar{\varepsilon}]$ and a $\delta > 0$ such that:

$$\gamma(A_c(x'),c) - \gamma(x',c) \leq -\delta$$

for all $x' \in N_\varepsilon(x)$. Hence any accumulation point x^* of an infinite sequence $\{x_i\}$, such that $x_{i+1} = A_c(x_i)$ for all i, lies in D_c.

It should be noted that the estimate H of L_{xx} does not affect convergence. A poor estimate may, of course, result in the solution of QP(x,H) yielding an unsatisfactory search direction; this will, however, be detected by the tests (in particular test (γ)) and the first order descent direction selected.

6. CONVERGENCE

Since $\bar{c}$ and A_c, for each algorithm, satisfy hypotheses (i)-(iii) of Theorem 1, we obtain:

Theorem 2

Let $\{x_i\}$ be a bounded infinite sequence generated by either algorithm. Then the penalty parameter is increased finitely often and any accumulation point x^* of $\{x_i\}$ lies in D. □

7. RATE OF CONVERGENCE

We now consider the rate of convergence of the second algorithm. We make the further assumption that at each Kuhn-Tucker triple $\{\hat{x},\hat{\lambda},\hat{\mu}\}$ for P_2, the second order sufficiency conditions hold with strict complementary slackness, i.e. $\hat{\lambda}^j > 0$ for all $j \in I(\hat{x})$ and $L_{xx}(\hat{x},\hat{\lambda},\hat{\mu})$ is positive definite on the subspace $\{p | g_x^j(\hat{x})p = 0, j \in I(\hat{x}); h_x(\hat{x})p = 0\}$. We hope to achieve superlinear convergence; this requires that H_i (the estimate of L_{xx} at iteration i) converges (in some sense) to the true value at a solution, that for all i sufficiently large, the tests (α), (β), (γ) are satisfied so that the second order search direction is selected, and that the Armijo step length rule yields a step length of unity for all i sufficiently large.

We do not make the conventional, but strong, assumption that $x_i \to x^*$ as $i \to \infty$ and that the second order search direction is selected for all i sufficiently large.

Let $\{x_i\}$ be an infinite, bounded, sequence generated by the algorithm, and let B be a compact ball in R^n containing $\{x_i\}$. Our assumptions imply that B contains a finite number of Kuhn-Tucker points for P_2; any accumulation point x^* of $\{x_i\}$ is a member of this set. Let K_1 denote the subsequence in which the first order search direction and K_2 the subsequence in which the second order search direction is employed. Since the penalty parameter remains constant at c for all i sufficiently large, it follows that $\bar{p}(x_i,c) \to 0$ and, hence, that $||x_{i+1} - x_i|| \to 0$ as $i \to \infty$, $i \in K_1$. If the test (β) for the second order search direction is replaced by:

$$(\beta) \qquad ||p|| \leq k\delta_i^j$$

where $\delta_1 \in (0,1)$, $k \in (0,\infty)$, and j is the number of times the second order search direction has been employed, then $||x_{i+1} - x_i|| \to 0$ as $i \to \infty$, $i \in K_2$. Since any accumulation point x^* of $\{x_i\}$ is a Kuhn-Tucker point, and since the number of Kuhn-Tucker points in B is finite, it follows finally that $x_i \to x^*$ as $i \to \infty$, where x^* is one of the Kuhn-Tucker points in B.

If H_i (the current estimate of L_{xx}) is obtained via a secant updating procedure (at teration i replace column i mod(n) of H_i by:

$$(1/\Delta_i)[\nabla_x h(x_{i+1} + \Delta_i e_i, \bar{\lambda}(x_{i+1}), \bar{\mu}(x_{i+1}) - \nabla_x h(x_{i+1}), \bar{\lambda}(x_{i+1}), \bar{\mu}(x_{i+1})]$$

where $\Delta_i = \min\{||x_{i+1} - x_i||, \varepsilon\})$ it follows from the continuity of $\bar{\lambda},\bar{\mu}$ that $H_i \to L_{xx}(x^*,\lambda^*,\mu^*)$ as $i \to \infty$ where (x^*,λ^*,μ^*) is a Kuhn-Tucker triple for P_2.

For each i let $\{p_i,\lambda_i,\mu_i\}$ denote the minimum norm Kuhn-Tucker triple for $QP(x_i,H_i)$. We can now employ the perturbation theory of Robinson (1974) to deduce that $(x_i,\lambda_i,\mu_i) \to (x^*,\lambda^*,\mu^*)$ as $i \to \infty$, that $QP(x_i,H_i)$ has a unique Kuhn-Tucker triple $\{p_i,\lambda_i,\mu_i\}$ for all i sufficiently large, $p_i \to 0$ as $i \to \infty$, and that $I_i \triangleq \{j | \lambda^j > 0\} = I(x^*)$ for all i sufficiently large. It follows that the test (α) is satisfied for all i sufficiently large.

Let the function T in test (γ) be defined by:

$$T(x) \triangleq \min\{\varepsilon,[\psi(x) + ||\nabla f(x) + g_x^T(x)\bar{\lambda}(x) + h_x^T(x)\bar{\mu}(x)||^2]^2\}$$

A detailed analysis, similar to that of Powell (1977), shows that $\Theta(x_i,p_i,c)$ is of order $-[T(x_i)]^{\frac{1}{2}}$ so that test (γ) is satisfied for all i sufficiently large. A standard analysis also shows that test (β) is satisfied for all i sufficiently large; satisfaction of this test follows from the superlinear convergence properties of this type of algorithm.

It remains therefore to establish that, for all i sufficiently large, the Armijo rule yields a step length of unity, i.e. the test:

$$\gamma(x_i+p_i,c) - \gamma(x_i,c) \leq \Theta(x_i,p_i,c)/8$$

is satisfied for all i sufficiently large (c denoting the constant value of c_i achieved when i is sufficiently large). Unfortunately this is not necessarily true as has been shown in the counter example provided by Maratos (1978). The problem is $\min\{||x||\,|(x^1+1)^2 + (x^2)^2 - 4 = 0,\ x \in R^2\}$ which has a solution at $x^* = (1,0)$ with multiplier $\lambda^* = -1/2$. Let x be such that $h(x) = 0$; a straightforward calculation shows that the second order search direction $p(x)$ (with $H = L_{xx}(x,\lambda(x)),\lambda(x) = (h_x(x)^T)^+\nabla f(x)$) satisfies

$$\Delta(x) \triangleq \gamma(x + p(x),c) - \gamma(x,c) - \Theta(x,p(x),c)/8$$

$$= 2(x^2)^2[2(c+1) - (7/8)(1+x^1)]/[1+(x^1)^2]$$

$$> 0$$

for all x satisfying $h(x) = 0$, $x^1 \in (-1,1)$ and for *all* $c > 0$; the step length is thus always less than one for such x. An analysis of the example reveals the source of the difficulty. It is easily verified that, although $\Theta(x,p(x),c)$ is negative, the change in ψ and also in γ, is positive. This can be avoided, at least to second order, if, as proposed in Maratos (1978), the search direction $p(x)$ is replaced by a search arc along which the second order estimate of ψ remains zero. Maratos employs the second order derivative of h at x to achieve this; to reduce computation we prefer to employ at x_i the search arc $\{\alpha p_i + \alpha^2\tilde{p}_i | \alpha \in [0,1]\}$ where $\tilde{p}_i$ is the minimum norm solution of:

$$g^j(x_i + p_i) + g_x^j(x_i)p_i = 0, \quad j \in I_i$$

$$h(x_i + \gamma_i) + h_x(x_i)p_i = 0$$

if a solution with norm not greater than $||p_i||$ exists, and is the zero vector otherwise. The complete algorithm can now be specified:

Algorithm 2

Data: $x_1 \in R^n$, b,c_0,δ, $k \in (0,\infty)$, $\delta_1 \in (0,1)$, $H_1 \in R^{n\times n}$

Step 0: Set $i = 1$, $j = 0$

Step 1: If $c_{i-1} \geq \bar{c}(x_i)$, set $c_i = c_{i-1}$

If $c_{i-1} < \bar{c}(x_i)$, set $c_i = \max\{c_{i-1}+\delta, \bar{c}(x_i)\}$

Step 2: *If:* (α) A (minimum norm) solution p_i of $QP(x_i,H_i)$ exists

(β) $||p_i|| \leq k\delta_1^j$

(γ) $\theta(x_i,\gamma_i,c_i) \leq -T(x_i)$

Then: (a) Compute $\tilde{p}_i$

(b) Compute α_i, the largest $\alpha \in \{1,\beta,\beta^2 ...\}$ such that:

$$\gamma(x_i + \alpha p_i + \alpha^2\tilde{p}_i,c_i) - \gamma(x_i,c_i) \leq \alpha\theta(x_i,p_i,c_i)/8$$

(c) Set $x_{i+1} = x_i + \alpha_i p_i + \alpha_i^2\tilde{p}_i$

(d) Update H_i to H_{i+1}

(e) Set i = 1+1, j = j+1 and go to Step 1.

Else: Proceed

Step 3: (a) Compute the first order descent direction $\bar{p}(x_i,c_i)$

(b) Compute α_i, the largest $\alpha \in \{1, \beta, \beta^2,...\}$ such that:

$$\gamma(x_i + \alpha\bar{p}(x_i,c_i),c_i) - \gamma(x_i,c_i) \leq \alpha\theta(x_i,\bar{p}(x_i,c_i),c_i)/4$$

(c) Set $x_{i+1} = x_i + \alpha_i\bar{p}(x_i,c_i)$

(d) Update H_i to H_{i+1}

(e) Set i = i+1 and go to Step 1. □

It can be shown that this modification leads to an asymptotic step length of unity and that $||\tilde{p}_i||$ is of order $||p_i||^2$ and does not destroy the superlinear convergence of the algorithm, yielding:

Theorem 3

If $\{x_i\}$ is a bounded infinite sequence generated by algorithm 2, then $x_i \to x^* \in D$ superlinearly. □

8. CONCLUSION

We have shown how exact penalty functions may be employed to enforce convergence of a first and second order algorithm for constrained optimization without destroying the superlinear rate of convergence of the second order algorithm. The performance of the first order algorithm appears satisfactory. The second order algorithm has the strong asymptotic properties stated in Theorem 3, but can be improved in several ways.

If H_i is not positive definite, $QP(x_i,H_i)$ may have more than one solution. The algorithm requires that the minimum norm Kuhn-Tucker triple for $QP(x_i,H_i)$ be determined, which is not computationally attractive. Powell (1977) overcomes this difficulty in an ingenious way; he exploits the fact that convergence of H_i to H^* is not required, merely the convergence of the projection of H_i (onto the tangent plane of the constraint manifold at x^*) to the projection of H^*, to employ an updating procedure which preserves the positive definiteness of H_i. We overcome the difficulty by adding to $QP(x_i,H_i)$ the constraint $||p_i|| \leq k\delta_i^j$. Because of the test (β), this does not change the algorithm, but automatically ensures selection of the minimum norm Kuhn-Tucker triple for all i sufficiently large.

Secondly, the test $c \geq \bar{c}(x_i)$ does not necessarily ensure that $\Theta(x_i,p_i,c) < 0$ if $x_i \notin D_c$ even if H_i is positive definite. This may result in the first order search direction being employed unnecessarily often. This effect can be reduced if $\bar{c}(x_i)$ is replaced by $\tilde{c}(x_i)$ defined by:

$$\tilde{c}(x_i) \triangleq \max \{\bar{c}(x_i), \sum \tilde{\lambda}^j + \sum|\tilde{\mu}^j|\}$$

where $\tilde{\lambda} = \lambda_i$ and $\tilde{\mu} = \mu_i$ if the Kuhn-Tucker triple $\{p_i,\lambda_i,\mu_i\}$ satisfies the test $||p_i|| < k\ \delta_1^j$ and $\tilde{\lambda} = 0$, $\tilde{\mu} = 0$ otherwise. This change does not affect the convergence properties of the algorithm.

Other changes can be made. If a solution to $QP(x_i,H_i)$ does not exist, the constraints may be relaxed, as proposed by Powell (1978), and the resultant search direction accepted if it satisfies the tests (α), (β) and (γ).

The procedure for updating the penalty parameter does not allow it to decrease. This may cause the parameter to be somewhat larger than necessary as a solution is approached. The asymptotic properties of the algorithm will not, however, be affected if the penalty parameter is decreased a finite number of times.

REFERENCES

1. Conn, A.R. "Constrained Optimization Using a Nondifferentiable Penalty Function", SIAM J. Numer. Anal. 10, 760-784, (1973).
2. Conn, A.R. and Pietrzykowski, T. "A Penalty Function Method Converging Directly to a Constrained Optimum", SIAM J. Numer. Anal. 14, 348-378, (1977).
3. Han, S.P. "A Globally Convergent Method for Nonlinear Programming Problems", JOTA, 22, 297-309, (1977).
4. Levitin, E.S. and Polyak, B.T. "Constrained Minimization Methods", USSR Computational Mathematics and Mathematical Physics, 6, 1-15, (1966).
5. Maratos, N. "Exact Penalty Function Algorithms for Finite Dimensional and Control Optimization Problems", Ph.D. thesis, Imperial College, London, (1978).
6. Mayne, D.Q. and Maratos, N. "A First Order Exact Penalty Function Algorithm for Equality Constrained Optimization Problems", Mathematical Programming, 16, 303-324, (1979).
7. Mayne, D.Q. and Polak, E. "A Superlinearly Convergent Algorithm for Constrained Optimization Problems", Research Report, C.C.D., Imperial College, 78/52, (1978).
8. Polak, E. "Computational Methods in Optimization, A Unified Approach", Academic Press, (1971).
9. Powell, M.J.D. "A Fast Algorithm for Nonlinearly Constrained Optimization Calculations", in Numerical Analysis, ed. G.A. Watson, Springer Verlag, (1978).

10. Powell, M.J.D. "The Convergence of Variable Metric Methods for Nonlinearly Constrained Optimization Problems", Technical Memorandum 315, Applied Mathematics Division, Argonne National Laboratory, Illinois, (1977).

11. Charalambous, C. "A lower bound for the controlling parameters of the exact penalty functions", Mathematical Programming, 15, 278-290, (1978).

12. Robinson, S.M. "Perturbed Kuhn-Tucker Points and Rates of Convergence for a Class of Nonlinear Programming Algorithms", Mathematical Programming, 7, 1-16, (1972).

PETROV-GALERKIN METHODS FOR NON-SELF-ADJOINT PROBLEMS

K.W. Morton

1. Introduction

For self-adjoint elliptic problems, finite element methods based on the variational or Galerkin formulation provide particularly appropriate and accurate approximations. If $a(u, v)$ is the symmetric positive-definite form associated with a second order elliptic operator L on a bounded region Ω, the variational form of the problem

$$Lu = f \quad \text{in} \quad \Omega, \qquad u = 0 \quad \text{on} \quad \partial\Omega, \tag{1.1}$$

becomes

$$a(u, v) = \langle f, v\rangle, \qquad \forall\ v \in H_0^1(\Omega), \tag{1.2}$$

where $\langle\cdot, \cdot\rangle$ denotes the L_2 inner product on Ω and $H_0^1(\Omega)$ is the Sobolev space of functions whose derivatives are L_2 integrable over Ω and which vanish on $\partial\Omega$. Suppose a finite element approximation takes the form

$$V(\underline{x}) = \sum_{(j)} Q_j\phi_j(\underline{x}), \qquad V = 0 \quad \text{on} \quad \partial\Omega \tag{1.3}$$

with the basis functions $\{\phi_j\}$ spanning a trial space $S_0^h \subset H_0^1(\Omega)$. Then the Galerkin approximation $U(\underline{x})$ given by

$$a(U, V) = \langle f, V\rangle, \qquad \forall\ V \in S_0^h, \tag{1.4}$$

has the key approximation property, in the energy norm,

$$a(u - U, u - U) = \min_{V\in S_0^h} a(u - V, u - V). \tag{1.5}$$

All the error estimates follow from this property as well as the confidence to use the most economical approximation that is capable of giving the required accuracy - an important counter advantage when comparing with the usually simpler finite difference methods. Moreover, (1.5) implies various superconvergence properties whose exploitation is an important feature of finite element methods.

Unfortunately, all these properties of the Galerkin method are lost when the problem (1.1) or (1.2) is not self-adjoint. The error in the energy norm is within a constant factor of the minimum error attainable with the trial space but the constant can be large. Thus the same asymptotic order of accuracy is retained but the practical results can be poor. Several authors [2, 4, 5, 7] have therefore considered the improvements that might be obtained with the more general Petrov-Galerkin method, in which a test space T_0^h, different from the trial space S_0^h, is introduced and (1.4) is replaced by

$$b(U, W) = \langle f, W\rangle, \qquad \forall\, W \in T_0^h, \tag{1.6}$$

where $b(u, w)$ is a coercive but unsymmetric bilinear form. A number of possibilities exist for the choice of the test spaces but Barrett & Morton [1] have explicitly taken as their aim the re-establishment of the optimal approximation property (1.5) for their approximation. Suppose a symmetrizing operator $N : H_0^1(\Omega) \to H_0^1(\Omega)$ can be found with the property that

$$b(v, Nw) \equiv b_E(v, w), \qquad \forall\, v \in H^1(\Omega), \qquad w \in H_0^1(\Omega), \tag{1.7}$$

where $b_E(v, w)$ is a symmetric, coercive form. Then we could take $T_0^h = NS_0^h$ and the corresponding Petrov-Galerkin equations become

$$b(U_N, N\phi_i) \equiv b_E(U_N, \phi_i) = \langle f, N\phi_i\rangle, \qquad \forall\, \phi_i \in S_0^h. \tag{1.8}$$

This implies immediately that

$$b_E(u - U_N, \phi_i) = 0,$$

i.e.

$$\|u - U_N\|_E = \min_{V\in S_0^h} \|u - V\|_E , \tag{1.9}$$

where $\|v\|_E^2 = b_E(v, v)$. For the one-dimensional diffusion-convection model problem

$$-u'' + ku = 0, \qquad u(0) = 0,\ u(1) = 0, \tag{1.10}$$

Barrett & Morton [1] found that exact symmetrization could only be achieved in a way which led to poorly conditioned equations for large $k\Delta x$. But an approximate symmetriser N_ε gave well-behaved equations and a solution U_{N_ε} for which one can establish the result

$$\|u - U_{N_\varepsilon}\|_E \le \|u - U^*\|_E + [16k^2/(k^2 + 8)]\|u - U^*\|_{E,I_\varepsilon} , \tag{1.11}$$

where U^* is the best fit to u in the $\|\cdot\|_E$ norm and the last norm denotes the average of the energy integrand over the interval in which the symmetriser is modified. With an appropriate choice of this interval one therefore obtains the estimate

$$\|u - U_{N_\varepsilon}\|_E \le 17\|u - U^*\|_E , \tag{1.12}$$

uniformly for all k and Δx. This is the type of result sought for singular perturbation problems and attained with the finite difference schemes of Il'in [6].

In practice, U_{N_ε} is extremely close to U^* even for more general problems than (1.10). This then allows optimal recovery techniques to be used to recover more accurate information about u; that is, phenomena of superconvergence can be established. For example, the thickness of the boundary layer in (1.10) can be calculated accurately even when it is a small fraction of the element size.

More details of these methods are presented at this conference in a contributed lecture. So we turn now to hyperbolic problems and a consideration of what Petrov-Galerkin methods might contribute to their accurate and stable approximation.

2. The scalar wave equation $u_t = au_x$

We adopt the viewpoint with evolutionary problems that the main objective of any numerical procedure is a complete description of the solution at a succession of discrete time instants, the detailed time-behaviour between these instants being of less concern. Thus we seek approximations of the form

$$U^n(x) = \sum_{(j)} U_j^n \phi_j(x) \tag{2.1}$$

at $t = t_n$, $n = 1, 2, \ldots$. Now suppose $U^n(x)$ is an optimal approximation to $u(x, t_n)$, the exact solution at the nth step. For a one-step scheme, the aim should then be to construct $U^{n+1}(x)$ as the corresponding optimal approximation to $u(x, t_{n+1})$ at the next step.

For Petrov-Galerkin methods and first order equations, optimality is most naturally interpreted in the least squares sense. The Galerkin formulation then has at least two properties to commend it. Firstly, if u satisfies the evolution equation $u_t = Lu$ and $(Lu)^A$ is given as an approximation to Lu, the best approximation $\dot{U}$ in the span of $\{\phi_i\}$ to the time derivative is given by

$$<\dot{U} - (Lu)^A, \phi_i> = 0, \qquad \forall \phi_i. \tag{2.2}$$

When $(Lu)^A$ is taken as LU, we obtain the familiar semi-discrete Galerkin approximation

$$<\dot{U} - LU, \phi_i> = 0, \qquad \forall \phi_i, \tag{2.3}$$

a system of ordinary differential equations which can be discretised with respect to time in various ways. With reasonably smooth solutions and small time steps this can be highly accurate: see Cullen & Morton [3] and references therein. (We leave until section 4 consideration of whether LU is a good approximation to Lu when U is given as an approximation to u.) The second important property of the Galerkin equations (2.3) concerns conservation. We can assume that $\{\phi_i\}$ spans the constant function, so that if L is such that a given quantity which is at most quadratic in u is conserved in the exact problem, this same conservation property holds for the approximation U: in particular,

$$\frac{d}{dt}\tfrac{1}{2}\|U\|^2 = <U, LU> = 0, \quad \text{if} \quad <u, Lu> = 0, \ \forall u. \tag{2.4}$$

However, for a given time discretisation the Galerkin scheme may be far from ideal. For example, if Euler time-stepping is applied to (2.3) the scheme is unstable unless $\Delta t = O(\Delta x^2)$. This is very restrictive and one may ask whether Petrov-Galerkin methods can redeem the situation. With the scalar wave equation and constant a, if $a\Delta t$ equals the uniform element length Δx, an optimal approximation U^n at time t_n should become the optimal approximation U^{n+1} at time $t_n + \Delta t$ by a simple shift of one element length. This is a standard property of finite difference schemes and is often called the unit CFL property. Morton & Parrott [8] have recently shown how this can be achieved by Petrov-Galerkin methods when $\{\phi_j\}$ corresponds to the piecewise linear approximation.

For $a > 0$, one requires that $U^{n+1}(x) = U^n(x + \Delta x)$ when $a\Delta t = \Delta x$ and hence that

$$<\sum_{(j)} (U^n_{j+1} - U^n_j)\phi_j - \Delta x \sum_{(j)} U^n_j \phi'_j,\ \chi_i> = 0 \tag{2.5}$$

for the test functions χ_i. This reduces to requiring that

$$<\phi_{j-1} - \phi_j - \Delta x \phi'_j,\ \chi_i> = 0, \qquad \forall\ i, j. \tag{2.6}$$

A particular choice, for the linear basis functions ϕ_j, is given by

$$\begin{aligned} \chi^+_j &= 4 - 6(x - j\Delta x)/\Delta x, \quad & x \in [j\Delta x, (j+1)\Delta x] \\ &= 0 \quad &, \quad x \notin [j\Delta x, (j+1)\Delta x]. \end{aligned} \tag{2.7}$$

Taking a linear combination of this function with ϕ_i as the test function gives the following scheme:

EPG Method

$$<U^{n+1} - (U^n + a\Delta t \partial_x U^n),\ (1 - \nu)\phi_i + \nu\chi^+_i> = 0, \tag{2.8}$$

with χ^+_i given by (2.7). We also have the following result which is proved in Morton & Parrott [8]:

Theorem. The EPG scheme (2.8) is stable for

$$0 \le a\Delta t/\Delta x \le \nu \le 1 \tag{2.9}$$

and it is second order accurate for $\nu = a\Delta t/\Delta x$.

We shall refer to the scheme with this choice of ν as EPGI. In terms of the nodal values, (2.8) takes the form

$$[1 + 1/6(1 - \nu)\delta^2](U^{n+1}_j - U^n_j) = \mu\Delta_0 U^n_j + \tfrac{1}{2}\mu\nu\delta^2 U^n_j, \tag{2.10}$$

where the difference operators have the usual meaning $\delta U_j = U_{j+\frac{1}{2}} - U_{j-\frac{1}{2}}$, $\Delta_0 U_j = \frac{1}{2}(U_{j+1} - U_{j-1})$ and $\mu = a\Delta t/\Delta x$. Thus with $\nu = \mu$, the EPGI scheme corresponds to the Lax-Wendroff method with a modified mass matrix. An analysis of the phase errors and amplification factors for a Fourier mode e^{ikx} shows that it has improved accuracy. Table I gathers together such data for all the schemes derived in this paper for the single mode given by $k\Delta x = \frac{1}{2}\pi$.

Similar methods can be derived for other time-stepping schemes. Thus the θ-scheme

$$<U^{n+1} - a\Delta t\theta\partial_x U^{n+1} - [U^n + a\Delta t(1 - \theta)\partial_x U^n],\ \psi_i> = 0 \tag{2.11}$$

satisfies the unit CFL property for $a > 0$ and the test function χ^+_i if this satisfies the relations

$$<\phi_{j-1} - \phi_j - \Delta x[\theta\phi'_{j-1} + (1-\theta)\phi'_j],\ \chi_i^+> = 0, \qquad \forall\ i,\ j. \tag{2.12}$$

This reduces to requiring that, if $\chi_j^+(x) = \chi(x/\Delta x - j)$ and $\chi(s) = 0$ if $s \notin [0, 1]$, then

$$\int_0^1 s\chi(s)ds = \theta \int_0^1 \chi(s)ds. \tag{2.13}$$

Euler time-stepping is included as the special case $\theta = 0$ and the particular choice corresponding to (2.7) is

$$\chi(s) = 2(2 - 3\theta) - 6(1 - 2\theta)s, \qquad s \in [0, 1]. \tag{2.14}$$

An important special case is Crank-Nicolson-Petrov-Galerkin for which $\chi_j^+(x)$ reduces to the characteristic function of the interval $[j\Delta x, (j+1)\Delta x]$. Then with $\psi_i = (1 - \nu)\phi_i + \nu\chi_i^+$ in (2.11) the nodal parameters satisfy

$$[1 + \tfrac{1}{2}\nu\Delta_0 + \frac{1}{6}(1 + \tfrac{1}{2}\nu)\delta^2](U_j^{n+1} - U_j^n) = \tfrac{1}{2}\mu[\Delta_0 + \tfrac{1}{2}\nu\delta^2](U_j^{n+1} + U_j^n). \tag{2.15}$$

The choice $\nu = 1$ leads to the familiar Keller "Box scheme", which like the Galerkin choice $\nu = 0$ is second order accurate. However, we can obtain a 3rd order accurate method with $\nu = \mu^2$:

CNPG Method: (2.15) with $\nu = \mu^2$.

Finally, the leapfrog time-stepping scheme

$$<U^{n+1} - U^{n-1} - 2a\Delta t\partial_x U^n,\ \psi_i> = 0 \tag{2.16}$$

satisfies the unit CFL property for a test function χ_i if

$$<\phi_{j-1} - \phi_{j+1} - 2\Delta x\phi'_j,\ \chi_i> = 0 \qquad \forall\ i,\ j. \tag{2.17}$$

Scheme	$\mu = 0.2$		$\mu = 0.55$		$\mu = 0.9$	
	Δ	κ	Δ	κ	Δ	κ
Lax-Wendroff	- .3462	0.675	- .2271	0.436	- .0360	0.715
EG	- .0722	2.367	- .2015	6.15	- .3398	7.96
EPGI	- .1060	0.724	- .0882	0.528	- .0107	0.797
EPGII	- .0358	0.755	.0058	0.641	.0122	0.886
LG	- .0301	1.0	.1229	1.0		
LPG	- .0497	1.0	- .0752	1.0	- .0874	1.0
Box	.2566	1.0	.1640	1.0	.0367	1.0
CNG	- .0521	1.0	- .0942	1.0	- .1600	1.0
CNPG	- .0424	0.944	- .0165	0.756	- .0100	0.888

Table I. Phase errors Δ in units of wave length and amplification factors κ after one oscillation for various methods applied to $u_t = au_x$, with $\mu = a\Delta t/\Delta x$ and $k\Delta x = \frac{1}{2}\pi$.

This requires that the generating function $\chi(s)$, $s \in [-1, 1]$, satisfies

$$\int_{-1}^{0} s\chi(s)ds = \int_{0}^{1} s\chi(s)ds = 0$$
$$\int_{-1}^{0} \chi(s)ds = \int_{0}^{1} \chi(s)ds, \tag{2.18}$$

the simplest choice being

$$\chi(s) = 2 - 3|s|, \qquad s \in [-1, 1], \tag{2.19}$$

independent of the sign of a. With $\psi_j = (1 - \nu)\phi_j + \chi_j$ the resulting scheme is stable for

$$3\mu^2 \leq 1 + 2\nu \leq 3. \tag{2.20}$$

The choice $\nu = \mu^2$ is 4th order accurate and stable for $\mu^2 \leq 1$. In terms of the nodal parameters we have:

<u>LPG Method</u>

$$[1 + \tfrac{1}{6}(1 - \mu^2)\delta^2](U_j^{n+1} - U_j^n) = 2\mu\Delta_0 U_j^n. \tag{2.21}$$

The results in Table I are enough to demonstrate how in all cases the Petrov-Galerkin methods extend the accuracy attained with Galerkin methods for small values of μ to the whole range $0 \leq \mu^2 \leq 1$. The phase accuracy is particularly high for CNPG, though there is some dissipation introduced, and the stability is improved for both EPGI and LPG.

To check whether these encouraging features survive for variable coefficient problems, where the choice of ν has to be a local one, Morton & Parrott [8] considered the test problem

$$\partial_t u + \partial_x\left(\frac{u}{1 + 2x}\right) = 0 \quad \text{on} \quad (0, \pi),$$
$$u(t = 0) = \sin 9x \quad \text{on} \quad (0, \tfrac{1}{3}\pi), \tag{2.22}$$

which has a known analytic solution. They used the simple choice

$$\nu_i = a(x_i)\Delta t/\Delta x \tag{2.23}$$

and the results were entirely consistent with the analysis for the constant co-efficient case.

3. Symmetric hyperbolic systems $\underline{u}_t = A\underline{u}_x$

Suppose the symmetric $m \times m$ matrix A is diagonalised by the orthogonal trans-form

$$A = S\Lambda S^T, \qquad \Lambda = \{\lambda_i\delta_{ij}\}, \tag{3.1}$$

and we set $\underline{u} = S\underline{v}$. Then we obtain the diagonalised form of the equations

$$\underline{v}_t = \Lambda\underline{v}_x \tag{3.2}$$

and can apply our schemes to this form. However, it will often be unacceptably inconvenient to carry out this transformation and we need to consider what the schemes will look like in the original variables. The EPG and CNPG schemes will be least convenient because of their dependence on the direction of the characteristics. So we must introduce the matrix formed from the $|\lambda_i|$,

$$\tilde{A} = S|\Lambda|S^T, \tag{3.3}$$

and the test functions for $\lambda_i < 0$, $\chi_j^-(x) = \chi(j - x/\Delta x)$ with χ given by (2.14) and the appropriate choice of θ. Then it is helpful to define

$$\sigma_j = \tfrac{1}{2}(\chi_j^+ + \chi_j^-) - \phi_j, \qquad \tau_j = \tfrac{1}{2}(\chi_j^+ - \chi_j^-) \tag{3.4}$$

and we obtain the Euler-Petrov-Galerkin equations which generalise EPGI in the form

$$\langle \underline{U}^{n+1} - (\underline{U}^n + \Delta t A\partial_x \underline{U}^n),\ \phi_i\, \underline{e}_{(r)}\rangle + (\Delta t/\Delta x)\langle \underline{U}^{n+1} - \underline{U}^n,\ \sigma_i \tilde{A}\underline{e}_{(r)}\rangle$$
$$= (\Delta t^2/\Delta x)\langle A\partial_x \underline{U}^n,\ \tau_i A\underline{e}_{(r)}\rangle, \qquad r = 1, 2, \ldots, m, \tag{3.5}$$

where $\underline{e}_{(r)}$ is the rth unit vector of R^m. The first term here is just the Euler-Galerkin approximation; the last is the second order correction corresponding to the Lax-Wendroff operator; and the middle term increases the diagonal dominance of the mass matrix because of the form of $\tilde{A}$. Note that if $\tilde{A}$ has to be approximated it is only the mass matrix that is affected.

We can eliminate even this problem, however, by modifying the test function to be used with the rth component of (3.2) to take the form

$$\begin{aligned}\psi_{j,r} &= (1 - \mu_r^2)\phi_j + \tfrac{1}{2}\mu_r^2\,(\chi_j^+ + \chi_j^-) + \tfrac{1}{2}\mu_r(\chi_j^+ - \chi_j^-)\\ &= \phi_j + \mu_r^2\sigma_j + \mu_r\tau_j,\end{aligned} \tag{3.6}$$

where $\mu_r = \lambda_r\Delta t/\Delta x$. Then we obtain a scheme which we call EPGII and which corresponds to (3.5) but with $\Delta t/\Delta x\tilde{A}$ replaced by $(\Delta t/\Delta x)^2A^2$. This is actually third order accurate and rivals CNPG in phase accuracy, though it is more dissipative (see Table I).

Generalisation of LPG to a system is more straightforward and it is readily seen that one obtains

$$\langle \underline{U}^{n+1} - \underline{U}^{n-1} - 2\Delta t A\partial_x U^n,\ \phi_i\underline{e}_{(r)}\rangle + \langle \underline{U}^{n+1} - \underline{U}^{n-1},\ (\Delta t/\Delta x)^2A^2(\chi_i - \phi_i)\underline{e}_{(r)}\rangle$$
$$= 0. \tag{3.7}$$

Preliminary tests on the wave equation, including reflection from a rigid wall, show greatly improved accuracy over the Galerkin methods in all cases: and LPG appears to be even more accurate than the staggered leap-frog scheme, while of course being more general.

4. Systems of conservation laws $\partial_t\underline{u} + \partial_x\underline{f}(\underline{u}) = 0$

Of the test functions generated from (2.14), the only set which span the constant function corresponds to $\theta = \tfrac{1}{2}$, the CNPG method. Thus this and the LPG scheme are the only ones which will reproduce a conservation property: CNPG is implicit so early studies have been confined to the LPG method. An obvious generalisation of (2.6) to a conservation law system is

$$<\underline{U}^{n+1} - \underline{U}^{n-1} + 2\Delta t \partial_x \underline{F}^n, \phi_i \underline{e}_{(r)}> + (\Delta t/\Delta x)^2 <J_i^2(\underline{U}^{n+1} - \underline{U}^{n-1}), (\chi_i - \phi_i)\underline{e}_r> = 0 \tag{4.1}$$

where $J_i = \partial \underline{f}/\partial \underline{\dot{u}}|_{\underline{u}=\underline{U}_i}$ and the order of the second term has had to be changed to cover the case when J_i is unsymmetric.

A severe test of such a scheme is provided by the shock tube problem used by Sod [9] in his survey article. With no damping or smoothing the scheme (4.1) gives rise to large amplitude oscillations, though the propagation speeds of the shock, contact discontinuity and rarefaction wave are well approximated. Such oscillations are common with most schemes for approximating shock problems and various forms of artificial viscosity or diffusion are used to damp them out. For the present class of methods alternative procedures suggest themselves, based on the ideas of optimal approximation which have guided progress so far.

Suppose $U(x)$ is the least squares best fit of a piecewise linear function to $u(x)$. Then we need to construct the similar best fit to $\partial_x f(u)$. The first step is to "recover" u from U. Where u is smooth we could for instance suppose it to be interpolated, to accuracy $O(\Delta x^4)$, by a cubic spline $U^{(C)}(x)$, entirely determined by its nodal parameters $U_j^{(C)}$. Thus we calculate a recovery cubic spline $U^{(R)}$ to approximate $U^{(C)}$ by requiring that U is its least squares best fit, that is,

$$<U^{(R)} - U, \phi_i> = 0, \qquad \forall \phi_i. \tag{4.2}$$

This gives a linear system of equations relating the nodal values $U_j^{(R)}$ to the given parameters U_j. In fact, it is only necessary to carry out this recovery operation locally, and one can establish the result:

Theorem. Suppose $u \in C^4(0, 1)$, and U is the least squares best fit of a piecewise linear function on a uniform mesh to u with $U(0) = u(0)$, $u(1) = u(1)$. Then

$$\left|u(j\Delta x) - (1 + \tfrac{1}{12}\delta^2)U_j\right| \le C(\Delta x)^4 \sup_x \left|u^{(4)}(x)\right|. \tag{4.3}$$

A similar recovery operation can be performed assuming that u is piecewise smooth with a number of discontinuities which are more than, say, four mesh widths apart. Thus from the assumption that $\underline{U}^n$ is the best fit to a shock wave $\underline{u}(x, t_n)$, we can recover a piecewise smooth function $\underline{\tilde{U}}^n$, with explicit information on shock positions and strengths, and use this to calculate $\underline{U}^{n+1}$ from, for example, the leapfrog scheme

$$<\underline{U}^{n+1} - \underline{U}^{n-1} - 2\Delta t \partial_x \underline{f}(\underline{\tilde{U}}^n), \psi_i> = 0. \tag{4.4}$$

The development of algorithms of this type is well advanced and appears to be highly promising. It constitutes a logical extension of the Two Stage Galerkin method described in Cullen & Morton [3] and shown there to be nearly six times more accurate than the usual Galerkin method for smooth flows.

References

[1] Barrett, J.W. & Morton, K.W. Optimal finite element solutions to diffusion-convection problems in one dimension. U. of Reading Num. Anal. Rep. 3/78 (197

[2] Christie, I, Griffiths, D.F., Mitchell, A.R. & Zienkiewicz, O.C. Finite element methods for second order differential equations with significant first derivatives. Int. J. Num. Meth. Eng. 10 (1976) 1389-1396.

[3] Cullen, M.J.P. & Morton, K.W. Analysis of evolutionary error in finite element and other methods. To appear in J. Comp. Phys. (1979).

[4] Heinrich, J.C., Huyakorn, P.S., Mitchell, A.R. & Zienkiewicz, O.C. An upwind finite element scheme for two-dimensional convective transport equations. Int. J. Num. Meth. Eng. 11 (1977) 131-143.

[5] Hemker, P.W. A numerical study of stiff two-point boundary problems. Thesis, Math. Cent. Amsterdam (1977).

[6] Il'in, A.M. Differencing scheme for a differential equation with a small parameter affecting the highest derivative. Math. Notes Acad. Sci. USSR 6 (1969) 596-602.

[7] Mock, M.S. Projection methods with different trial and test spaces. Math. Comp. 30 (1976) 400-416.

[8] Morton, K.W. & Parrott, A.K. Generalised Galerkin methods for first order hyperbolic equations. U. of Reading Num. Anal. Rep. 4/78 (1978).

[9] Sod, G.A. A survey of several finite difference methods for systems of non-linear hyperbolic conservation laws. J. Comp. Phys. 27 (1978) 1-31.

COLLOCATION AND PERTURBED COLLOCATION METHODS

SYVERT P. NØRSETT

1. INTRODUCTION.

The problem to be solved is the initial value problem

(1.1) $$y' = f(t,y) \quad , \quad y(t_0) = y_0 \quad , \quad y \in \mathbf{R}^n \quad .$$

Among the methods used for solving (1.1) numerically are the Runge-Kutta methods (RK-methods),

(1.2)
$$k_i = f\left(t_0+c_ih,u_0+h\sum_{j=1}^{m} a_{ij}k_j\right) \quad , \quad i = 1,\cdots,m$$
$$u_1 = u_0+h\sum_{i=1}^{m} b_ik_i \quad , \quad u_i \sim y_i = y(t_i) \quad , \quad t_i = t_0+ih \quad .$$

The construction of methods of this form, of a given order, is not always easy. However, based on the work of BUTCHER [2], HAIRER and WANNER [8], [9] introduced the concept of Butcher series and this simplified the construction considerably.

Some of these RK-methods may on the other hand be obtained in a direct way by *collocation*. This fact was observed by WRIGHT [15], but he was not able to give a general result on the order of the collocation methods. This orderresult was pointed out by HULME [10], [11], but the proof he gave was not correct. A very short proof was given by NØRSETT and WANNER [14]. Independently and much earlier a nice proof was also presented by GUILLON and SOULÉ [7].

In BURRAGE [1] a class of RK-methods, called transformed methods was given. They are not collocation methods, but in some sense they are very close to be of that form. But by perturbing the collocation idea, one can include this class in the class of perturbed collocation methods (PECO-methods). Further, by choosing the pertubation operator suitably one is indeed able to give an equivalence between the class of RK-methods and the class of PECO-methods, NØRSETT and WANNER [13]. Order conditions for the PECO-methods are some more difficult to obtain than for ordinary collocation methods, but still easy to check.

In section 2 of this paper we give a short review of the collocation results and in section 3 we introduce the PECO-methods and list some of the results obtained by NØRSETT AND WANNER [13].

Section 4 is devoted to some comments and results on extensions of the collocation- and the perturbed collocation idea to multistep collocation for (1.1) and to collocation and perturbed collocation for the second order problem

$$(1.2) \qquad y'' = f(t,y,y') \quad , \quad y(t_0) = y_0 \quad , \quad y'(t_0) = y_0 \quad , \quad y \in \mathbf{R}^n .$$

2. COLLOCATION.

Fix $c_1,\cdots,c_m$ in $\mathbf{R}$, $c_i \neq c_j$ when $i \neq j$. The collocation method is defined by;

Find $u \in \pi_m$ such that

$$(2.1) \qquad u(t_0) = y_0$$

and

$$(2.2) \qquad u'(t_0+c_ih) = f(t_0+c_ih, u(t_0+c_ih)) \quad , \quad i = 1,\cdots,m \quad .$$

Then

$$(2.3) \qquad u_1 = u(t_0+h) \quad .$$

Define the Lagrange fundamental polynomials $\ell_i(\tau)$ by

$$\ell_i(\tau) = \prod_{\substack{j=1 \\ j\neq i}}^{m} \frac{\tau - c_j}{c_i - c_j} \quad ,$$

and set $k_i = u'(t_0+c_ih)$. The collocation method can then be written as a RK-method with

$$(2.4) \qquad a_{ij} = \int_0^{c_i} \ell_j(\tau)d\tau \quad , \quad b_i = \int_0^1 \ell_i(\tau)d\tau \quad .$$

Let y be the solution of (1.1) and u the solution of

$$u'(t) = f(t,u(t))+\delta(t) \ , \quad u(t_0) = y_0 \ .$$

Then the Theorem of GRÖBNER and ALEKSEEV says

$$(2.5) \qquad u(t)-y(t) = \int_{t_0}^{t} \Phi(t;\tau,u(\tau))\delta(\tau)d\tau$$

where Φ is a variational matrix. A short proof of this fundamental result is given in NØRSETT and WANNER [14]. When u is the collocation polynomial and

$$\delta(t) = u'(t)-f(t,u(t))$$

we have $\delta(t_0+c_ih) = 0$, $i = 1,\cdots,m$. A simple use of numerical quadrature then yields

THEOREM 1. The collocation method (2.1) - (2.3) has order $p \geq m$ iff the quadrature formula based on $c_1,\cdots,c_m$ has order p.

EXAMPLE 2. For m = 2 the corresponding RK-method is given by the matrix

$$\tilde{A} = \begin{bmatrix} \frac{-c_1(c_1-2c_2)}{2(c_2-c_1)} & -\frac{c_1^2}{2(c_2-c_1)} & 0 \\ \frac{c_2^2}{2(c_2-c_1)} & \frac{c_2(c_2-2c_1)}{2(c_2-c_1)} & 0 \\ \frac{2c_2-1}{2(c_2-c_1)} & \frac{1-2c_1}{2(c_2-c_1)} & 0 \end{bmatrix} .$$

REMARKS. (1) Explicit methods are not possible unless $c_1 = c_2 = 0$.

(2) The Ehle I_A-method of order 3 (Ehle [6]) is given by

$$\tilde{A} = \begin{bmatrix} \frac{1}{4} & -\frac{1}{4} & 0 \\ \frac{1}{4} & \frac{5}{12} & 0 \\ \frac{1}{4} & \frac{3}{4} & 0 \end{bmatrix}$$

whereas the collocation method based on $c_1 = 0$, $c_2 = \frac{2}{3}$ has

$$\tilde{A} = \begin{bmatrix} 0 & 0 & 0 \\ \frac{1}{3} & \frac{1}{3} & 0 \\ \frac{1}{4} & \frac{3}{4} & 0 \end{bmatrix}$$

and this is the Butcher I-method of order 3. Hence not all implicit RK-methods are covered by the collocation approach.

3. PERTURBED COLLOCATION.

It is easily shown that the matrix $A = \{a_{ij}\}_{i,j=1}^{m}$ satisfies $C(m)$ where

$$C(s) = \sum_{j=1}^{m} a_{ij} c_j^{k-1} = \frac{c_i^k}{k} \quad , \quad i = 1,\cdots,m \ ; \ k \leqslant s \quad .$$

When the order is m and $C(m)$ is true, the error at t_0+c_ih is at least of order m. But the transformed methods of Burrage satisfy $C(m-1)$ only, meaning that the error at t_0+c_ih is at least $m-1$ when the order is m. We therefore have to modify the collocation idea such that when the solution y is in π_r, $r \leqslant m$ the modification of (2.2) and (2.2) itself are identical. Let $N_j \in \pi_m$,

$$(3.1) \qquad N_j(t) = \frac{1}{j!} \sum_{i=0}^{m} (p_{ij} - \delta_{ij}) t^i, \quad \delta_{ij} \text{ the Kroneckerdelta.}$$

The *perturbation operator* $P : \pi_m \rightarrow \pi_m$ is defined by

$$(3.2) \qquad (Pu)(t) := u(t) + \sum_{j=1}^{m} N_j\left(\frac{t-t_0}{h}\right) u_0^{(j)} h^j \quad , \quad u_0^{(j)} = u^{(j)}(t_0) \quad .$$

Since P is linear, $j!N_j(x) = (Px^j)(x) - x^j$.

DEFINITION 3. For given P the perturbed collocation method is defined as for collocation except that (2.2) is replaced by

$$(3.3) \qquad u'(t_0+c_ih) = f(t_0+c_ih, (Pu)(t_0+c_ih)) \quad , \ i = 1,\cdots,m \ .$$

REMARK 4. To each N_j we can add an arbitrary multiple of the *collocation polynomial* M(t),

$$(3.4) \qquad M(t) = \prod_{i=1}^{m} (t-c_i)$$

without changing (3.3). Hence we assume that $N_m^{(m)} = -1$, $N_j \in \pi_{m-1}$; i.e. $P : \pi_m \to \pi_{m-1}$.

By proceeding as for the collocation method we find that the PECO-method is a RK-method with

$$(3.5) \qquad a_{ij} = \int_0^{c_i} \ell_j(\tau)d\tau + \sum_{\alpha=1}^{m} N_\alpha(c_i)\ell_j^{(\alpha-1)}(0)$$

and b_i as in (2.4). (Observe the perturbation of the a_{ij}-value from collocation.)

EXAMPLE 5. Let m = 2, $N_1(t) = 0$ and $N_2(t) = -\frac{t^2}{2} + at + b$. Using (3.5) we get

$$a_{11} = -\frac{c_1(c_1-2c_2)}{2(c_2-c_1)} - \frac{N_2(c_1)}{c_2-c_1}$$

$$a_{12} = -\frac{c_1^2}{2(c_2-c_1)} + \frac{N_2(c_1)}{c_2-c_1}$$

$$a_{21} = \frac{c_2^2}{2(c_2-c_1)} - \frac{N_2(c_2)}{c_2-c_1}$$

$$a_{22} = \frac{c_2(c_2-2c_1)}{2(c_2-c_1)} + \frac{N_2(c_2)}{c_2-c_1} .$$

With $c_1 = 0$ and $N_2(t) = -\frac{1}{2}t^2$ the following RK-method results,

$$\tilde{A} = \begin{bmatrix} 0 & 0 & 0 \\ c_2 & 0 & 0 \\ 1-\frac{1}{2c_2} & \frac{1}{2c_2} & 0 \end{bmatrix} .$$

This is the 2-stage explicit RK-method of order 2. With $c_1 = 0$, $c_2 = \frac{2}{3}$,

$$\tilde{A} = \begin{bmatrix} -\frac{3}{2}b & \frac{3}{2}b & 0 \\ \frac{2}{3} - a - \frac{3}{2}b & a + \frac{3}{2}b & 0 \\ \frac{1}{4} & \frac{3}{4} & 0 \end{bmatrix} .$$

In particular: $b = 0$ and $a = \frac{1}{3}$ give the Butcher I-method of order 3.
$b = -\frac{1}{6}$ and $a = \frac{2}{3}$ give the Ehle I_A-method of order 3.

We have observed that the PECO-methods are included in the set of RK-methods, but they have a much greater freedom than the collocation methods. The operator P is free to choose. Let the m×m-matrix $\tilde{P}$ be defined by

(3.6)
$$\tilde{P} = \begin{bmatrix} p_{01} & \cdots & p_{0m} \\ \vdots & & \\ p_{m-1\,1} & \cdots & p_{m-1\,m} \end{bmatrix} .$$
$$\underbrace{\phantom{p_{m-1\,1}}}_{N_1(t)} \qquad \underbrace{\phantom{p_{m-1\,m}}}_{N_m(t)}$$

Then $\tilde{P}$ is a matrix representation of P. From NØRSETT and WANNER [13] we have,

(3.7)
$$\tilde{A} = V\tilde{P}JV^{-1} = VA_PV^{-1}, \quad J = \text{diag}\left\{1, \frac{1}{2}, \cdots, \frac{1}{m}\right\}, \quad V = \left\{c_i^{j-1}\right\}_{i,j=1}^{m},$$

and

(3.8)
$$b^T = [b_1, \cdots, b_m]^T = [1, \cdots, 1]JV^{-1} .$$

When a RK-method has b as in (3.8) we call it *interpolatory*. In other words,

THEOREM 6. Each interpolatory RK-method with distinct nodes is equivalent to a PECO-method.

EXAMPLE 7. With $N_i(t) = 0$, $i = 1, \cdots, m-1$ and $N_m(t) = -\frac{1}{m!}t^m + \sum_{i=0}^{m-1} \alpha_i t^i$ the transformed methods of Burrage appear. In particular $N_m(t) = -\frac{1}{m!}M(t)$ gives the collocation methods.

Based on the nonlinear variation of constants formula of GRÖBNER and

ALEKSEEV (2.5), NØRSETT and WANNER [13] have shown,

THEOREM 8. Suppose that, for given $r \leq m$, $s \geq 0$

1) $N_j \equiv 0$ for $j = 1,\cdots,r-1$

2) $N_j \in \pi_j$

3) $\int_0^1 x^{k-1} N_j(x)\,dx = 0$, $j = r,\cdots,m$; $k = 1,\cdots,m+s-j$

4) $\int_0^1 x^{k-1} M(x)\,dx = 0$, $k = 1,\cdots,s$

5) $2r \geq m+s$.

Then the order of the corresponding PECO-method is at least m+s.

REMARKS.

i) For 1) $\operatorname{Ker}(P-I) = \pi_{r-1}$, and the order is at least r-1.

ii) For the error term in the quadrature formula we need that $u_0^{(j)}$, $j \leq m$, are *uniformly bounded* as $h \to 0$. By requiring $N_j \in \pi_j$ this can be assured.

The 4-th order method of Kutta is given by

$$\tilde{A} = \begin{bmatrix} 0 & 0 & 0 & 0 \\ \frac{1}{3} & 0 & 0 & 0 \\ -\frac{1}{3} & 1 & 0 & 0 \\ 1 & -1 & 1 & 0 \end{bmatrix} \quad \text{and} \quad \tilde{P} = \begin{bmatrix} 0 & 0 & 0 & 0 \\ 1 & -\frac{7}{3} & -\frac{1}{2} & \frac{10}{27} \\ 0 & 9 & \frac{3}{2} & -2 \\ 0 & -6 & 0 & \frac{8}{3} \end{bmatrix}.$$

Applied at $y' = y$, $y(0) = y_0$ we get

$$u(t) = \left[1+t+\tfrac{1}{2}\left(1-\tfrac{7}{6}h+\tfrac{1}{3}h^2\right)t^2+\tfrac{1}{3}\left(\tfrac{9}{2}-\tfrac{3}{2}h\right)t^3+\tfrac{1}{4}\left(\tfrac{3}{2}-\tfrac{3}{h}\right)t^4\right]y_0 .$$

In particular, $u^{(4)}(t) = 18\left(\frac{1}{2}-\frac{1}{h}\right) \to \infty$ when $h \to 0$.

Observe that $N_2(t) = -\frac{7}{6}t+4t^2-3t^3 \notin \pi_2$.

iii) The conditions in 3) are quadrature conditions for N_j and those in 4) are quadrature conditions for M. For the Kutta-method we have

$$\int_0^1 N_3(t)dt = \int_0^1 N_2(t)dt = 0 \ , \quad \int_0^1 tN_2(t)dt = \frac{1}{90} \ ,$$

and 3) is violated.

iv) Due to nonlinearity we need that $O(h^{2r}) = O(h^{m+s})$.

EXAMPLE 9. Let r = m-1, s = m-2. By Theorem 8,

$$M(t) = \binom{2m}{m}^{-1}\left[P_m(t)+\alpha_1 P_{m-1}(t)+\alpha_2 P_{m-2}(t)\right]$$

$$N_m(t) = -\left[\binom{2m}{m}m!\right]^{-1}\left[P_m(t)+\beta_1 P_{m-1}(t)+\beta_2 P_{m-2}(t)\right]$$

$$N_{m-1}(t) = \left[\binom{2m}{m}m!\right]^{-1}\left[\gamma P_{m-1}(t)\right]$$

where P_m is the Legendre polynomial,

$$P_m(t) = \sum_{j=0}^{m} (-1)^{m+j}\binom{m}{j}\binom{m+j}{j}t^j \ .$$

When the method of this example is applied at the test equation $y' = \lambda y$, $\lambda = zh$, we get

$$u_1 = \left\{\sum_{i=0}^{m} N^{(i)}(1)z^{m-i} \Big/ \sum_{i=0}^{m} N^{(i)}(0)z^{m-i}\right\}u_0 \tag{3.7}$$

where N is the *N-polynomial* (NØRSETT and WANNER [13]),

$$N(t) = \left[1+N_{m-1}^{(m-1)}(0)\right]N_m(t)-N_m^{(m-1)}(0)N_{m-1}(t)+\int_0^t N_{m-1}(x)dx \ . \tag{3.8}$$

From NØRSETT and WANNER [13],

THEOREM 10. The method of Example 9 is A-stable iff

$$\beta_1+\gamma_1/2 \leq 0$$

$$(\beta_2+1)\left(1 + \frac{\gamma_1}{4m-2}\right) \leq 0. \qquad ¤$$

Based on these results the following interesting table follows,

Method	α_1	α_2	β_1	β_2	γ	Colloc.	A-stable
Butcher I, [4]	1	0	1	0	0	Yes	No
Ehle I_A, [6]	1	0	-1	0	0	No	Yes
Butcher II, [4]	-1	0	1	0	0	No	No
Ehle I_A, [6]	-1	0	-1	0	0	Yes	Yes
Butcher III, [4]	0	-1	$\frac{2m-1}{m-1}$	$\frac{m}{m-1}$	0	No	No
Ehle III_A, [6]	0	-1	0	-1	0	Yes	Yes
Ehle III_B, [6]	0	-1	2m-1	$\frac{m}{m-1}$	-4m+2	No	Yes
Ehle-Chipman III_C, [5]	0	-1	$-\frac{2m-1}{m-1}$	$\frac{m}{m-1}$	0	No	Yes

In NØRSETT and WANNER [13], the *B-stability* of the methods of Example 9 is also considered. As a special result we have that the methods of order at least 2m-1, i.e. $\alpha_2 = \beta_2 = \gamma_1 = 0$ are B-stable iff $\beta_1 \leq 0$. Hence α_1 is free to choose.

4. EXTENSIONS.

In the collocation method we can increase the order by adding more collocationpoints. However the cost is more work when solving for the k_i-values. Instead of increasing the number of collocationpoints one can put in backinformation.

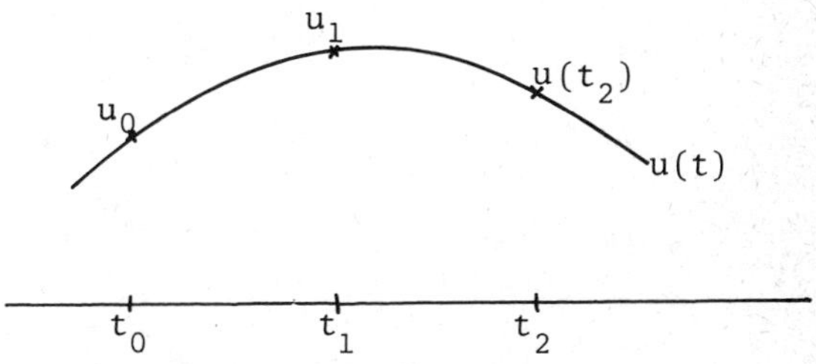

With u_0, u_1 given we define *Two-step collocation* as

Find $u \in \pi_{m+1}$ such that

(4.1) $\qquad u(t_i) = u_i \quad , \quad i = 0,1$

(4.2) $\qquad u'(t_0+c_ih) = f(t_0+c_ih, u(t_0+c_ih)) \quad , \quad i = 1,\cdots,m,$

giving $u_2 = u(t_2)$. The extension to k-step collocation is obvious and also backinformation from $f_0, f_1, \cdots, f_{k-1}$ can be added. For similar ideas, see LINK [12].

In order to study the order of the two-step collocation method we assume that $u_0 = y_0$, $u_1 = y_1$. From (2.5),

$$\int_{t_0}^{t_2} \Phi(t_2;\tau,u(\tau))\delta(\tau)d\tau = \int_{t_1}^{t_2} \Phi(t_2;\tau,u(\tau))\delta(\tau)d\tau$$

or

$$\int_{t_0}^{t_1} \Phi(t_2;\tau,u(\tau))\delta(\tau)d\tau = 0 \quad .$$

Hence for every real γ,

$$(4.3) \qquad \begin{aligned} u_2-y_2 &= \int_{t_0}^{t_2} \Phi(t_2;\tau,u(\tau))\delta(\tau)d\tau - \gamma \int_{t_0}^{t_1} \Phi(t_2;\tau,u(\tau))\delta(\tau)d\tau \\ &= \int_{t_0}^{t_2} \Phi(t_2;\tau,u(\tau))\kappa(\tau)\delta(\tau)d\tau \end{aligned}$$

where

$$\kappa(\tau) = \begin{cases} 1-\gamma, & t_0 \leqslant \tau \leqslant t_1 \\ 1 \;, & t_1 \leqslant \tau \leqslant t_2 \end{cases} \quad .$$

Since we seek $u \in \pi_{m+1}$, $u_2 - y_2 = \mathcal{O}(h^{m+2})$ for all $c_1, \cdots, c_m$. This is obtained from (4.3) if

$$(4.4) \qquad \gamma = \int_0^2 M(\tau)d\tau \Big/ \int_0^1 M(\tau)d\tau \quad .$$

The general result is,

THEOREM 11. The two-step collocation method has order m+1+s iff

$$(4.5) \qquad \int_0^2 M(\tau)\tau^i d\tau = \gamma \int_0^1 M(\tau)\tau^i d\tau \quad , \quad 1 \leqslant i \leqslant s$$

with γ given by (4.4) □

Another important result is,

THEOREM 12. The two-step collocation method is zero-stable (D-stable) iff

$$0 \leqslant \gamma < 2. \tag{4.6}$$ □

EXAMPLE 13. For m = 1 we get,

$$(2c_1-1)u_2-4(c_1-1)u_1+(2c_1-3)u_0 = 2hk_1 \quad ,$$

$$k_1 = f\left(t_0+c_1h,\ \frac{1}{2c_1-1}\left[-(c_1-1)^2u_0+c_1^2u_1+c_1(c_1-1)hk_1\right]\right) .$$

When $c_1 = 1$ we have the midpoint rule, $u_2-u_0 = 2hf_1$. Setting $c_1 = \frac{3}{2}$ the method is of Adams-type,

$$u_2-u_1 = hk_1$$

$$k_1 = f\left(t_0+\frac{3}{2}h,\ -\frac{1}{8}u_0+\frac{9}{8}u_1+\frac{3}{8}hk_1\right) .$$

The maximal order 3 is obtained for $c_1 = \frac{3+\sqrt{3}}{3}$ since we need $c_1 \geqslant 1$ for zero-stability. Further A_0-stability is assured if $c_1 \geqslant 1+\frac{\sqrt{2}}{2}$.

EXAMPLE 14. With m = 2, $c_1 = 0$ and $c_2 = 1$ we obtain zero-stability if $\frac{3}{2} < c_3 \leqslant 2$ and the order is 5 when $c_3 = \frac{11+\sqrt{41}}{10} = 1.740312424$. For arbitrary c_3 the methods are,

$$k_3 = f\left(t_0+c_3h,\left[1+\frac{c_3^3(c_3-2)}{2c_3-1}\right]u_0-\frac{c_3^3(c_3-2)}{2c_3-1}\,u_1\right.$$

$$\left.+\frac{c_3(c_3-1)}{2(2c_3-1)}h\left[(c_3-1)f_0+c_3^2f_1+k_3\right]\right)$$

$$(2c_3-1)u_2-8(2-c_3)u_1+(17-10c_3)u_0$$

$$= 2h\left[\frac{(2c_3^2-4c_3+1)}{c_3}f_0+\frac{(4c_3^2-9c_3+4)}{c_3-1}f_1+\frac{1}{c_3(c_3-1)}k_3\right] .$$

Observe that $k_1 = f_0$, $k_2 = f_1$. With $c_3 = 2$ we obtain the Milne Simpson formula.

The *perturbed two-step collocation* is now obvious. Replace (4.2) with

(4.2') $$u'(t_0+c_ih) = f(t_0+c_ih,(Pu)(t_0+c_ih)) \quad , \quad i = 1,\cdots,m$$

where P is a linear operator from π_{m+1} to π_{m+1}.

For the *second order problem* (1.2) the collocation method amounts to find $u \in \pi_{m+1}$ such that

(4.6) $$u(t_0) = u_0 \quad , \quad u'(t_0) = u'_0$$

(4.7) $$u''(t_0+c_ih) = f(t_0+c_ih,u(t_0+c_ih),u'(t_0+c_ih)), \ i = 1,\cdots,m$$

and

$$u_1 = u(t_0+h) \quad , \quad u'_1 = u'(t_0+h) \quad .$$

With $k_i = u''(t_0+c_ih)$ the method can be written as the implicit Nyström-method

$$k_i = f\left(t_0+c_ih,u_0+c_ihu'_0+h^2\sum_{j=1}^{m} a_{ij}k_j,u'_0+h\sum_{j=1}^{m}\bar{a}_{ij}k_j\right) \quad , \ i = 1,\cdots,m \ ,$$

(4.8) $$u'_1 = u'_0+h\sum_{i=1}^{m}\bar{b}_ik_i$$

$$u_1 = u_0+hu'_0+h^2\sum_{i=1}^{m} b_ik_i$$

where

(4.9)
$$a_{ij} = \int_0^{c_i}(c_i-\tau)\ell_j(\tau)d\tau \quad , \quad \bar{a}_{ij} = \int_0^{c_i}\ell_j(\tau)d\tau$$

$$b_i = \int_0^1(1-\tau)\ell_i(\tau)d\tau \quad , \quad \bar{b}_i = \int_0^1 \ell_i(\tau)d\tau \ .$$

By using the Gröbner-Alekseev result (2.5) we find,

$$u'(t)-y'(t) = \int_{t_0}^{t} \Lambda(t;\tau,u(\tau),u'(\tau))\delta(\tau)d\tau$$

$$u(t)-y(t) = \int_{t_0}^{t} (t-\tau)\Gamma(t;\tau,u(\tau),u'(\tau))\delta(\tau)d\tau$$

with

$$\delta(\tau) = u''(\tau)-f(\tau,u(\tau),u'(\tau))$$

from which the order conditions can be derived.

To get the *perturbed* collocation method we replace (4.7) with

$$u''(t_0+c_ih) = f\Big(t_0+c_ih,(Qu)(t_0+c_ih),\ (Pu')(t_0+c_ih)\Big)\ ,\ i = 1,\cdots,m \tag{4.7'}$$

where P,Q are linear operators, $P : \pi_m \to \pi_m$ and $Q : \pi_{m+1} \to \pi_{m+1}$.
When f is independent of y' we can, by elimination of u_0', assuming u_1 known and setting $u_2 = u(t_2)$, obtain the following two-step version,

$$\begin{aligned} k_i &= f\Big(t_0+c_ih,(1-c_i)u_0+c_iu_1+h^2\sum_{j=1}^{m}\hat{a}_{ij}k_j\Big)\ ,\quad i = 1,\cdots,m \\ u_2-2u_1+u_0 &= h^2\sum_{i=1}^{m}\hat{b}_ik_i \end{aligned} \tag{4.10}$$

$$\begin{aligned} \hat{a}_{ij} &= \int_0^{c_i}(c_i-\tau)\ell_j(\tau)d\tau-c_i\int_0^1(1-\tau)\ell_j(\tau)d\tau \\ \hat{b}_i &= \int_0^2(2-\tau)\ell_i(\tau)d\tau-2\int_0^1(1-\tau)\ell_i(\tau)d\tau\ . \end{aligned} \tag{4.11}$$

EXAMPLE 15. For m = 1 in (4.10) we have

$$k_1 = f\Big(t_0+c_1h,(1-c_1)u_0+c_1u_1+\frac{h^2}{2}c_1(c_1-1)k_1\Big)$$

$$u_2-2u_1+u_0 = h^2k_1\ .$$

With $c_1 = 1$, $k_1 = f_1$.

REFERENCES.

[1] BURRAGE K.; "A special family of Runge-Kutta methods for solving stiff differential equations". BIT 18 (1978) 22-41.
[2] BUTCHER J.C.; "Coefficients for the study of Runge-Kutta integration processes". J.Austral. Math. Soc. 3 (1963) 185-201.
[3] BUTCHER J.C.; "Implicit Runge-Kutta processes". Math. Comp. 18 (1964) 50-64.
[4] BUTCHER J.C.; "Integration processes based on Radau quadrature formulas". Math. Comp. 18 (1964) 233-243.
[5] CHIPMAN F.H.; "A-stable Runge-Kutta Processes". BIT 11 (1971) 384-388.
[6] EHLE B.L.; "High order A-stable methods for the numerical solution of D.E.'s". BIT 8 (1968) 276-278.
[7] GUILLON A., SOULE F.L.; "La resolution numerique des problèmes differentielles aux conditions initials par des méthodes de collocation." R.A.I.R.O. 3 (1969) 17-44.
[8] HAIRER E., WANNER G.; "Multistep-multistage-multiderivative methods and ordinary diff.eq". Computing 11 (1973) 287-303.
[9] HAIRER E., WANNER G.; "On the Butcher group and general multivalue methods". Computing 13 (1974) 1-15.
[10] HULME B.L.; "One-step piecewise polynomial Galerkin methods for initial value methods". Math. Comp. 26 (1972) 415-426.
[11] HULME B.L.; "Discrete Galerkin and related one-step methods for ordinary differential equations". Math. Comp. 26 (1972) 881-891.
[12] LINK B.D.; "Numerical solution of stiff. o.d.e.'s using collocation methods". Rep. 76-813, Dept. of Comp. Science, Univ. of Illinois, U.S.A.
[13] NØRSETT S.P., WANNER G.; "Perturbed collocation and Runge-Kutta methods". Submitted Num. math.
[14] NØRSETT S.P., WANNER G.; "The real-pole sandwich for rational approximations and oscillation equations". BIT 19 (1979) 79-94.
[15] WRIGHT K.; "Some relationships between implicit Runge-Kutta, collocation and Lanczos τ-method, and their stability properties". BIT 10 (1970) 217-227.

MODULAR ANALYSIS OF NUMERICAL SOFTWARE

Hans J. Stetter

1. INTRODUCTION

The June '79 issue of the Transactions on Mathematical Software (TOMS) contained an article by H. Crowder, R.S. Dembo, and J.M. Mulvey "On reporting computational experiments with mathematical software". The "Algorithms Policy" of TOMS is explained every few issues. But beyond that, the reporting (on scientific level) about *any* aspect of numerical software meets with peculiar difficulties as is well-known to every author who has written papers on numerical software and to every editor who has had to decide about the publication of such papers.

One of the reasons for this may derive from the fact that nearly all of these authors and editors have been trained as mathematicians while the design and evaluation of mathematical software is largely an *engineering activity* and thus outside their realm of experience.

Fig. 1

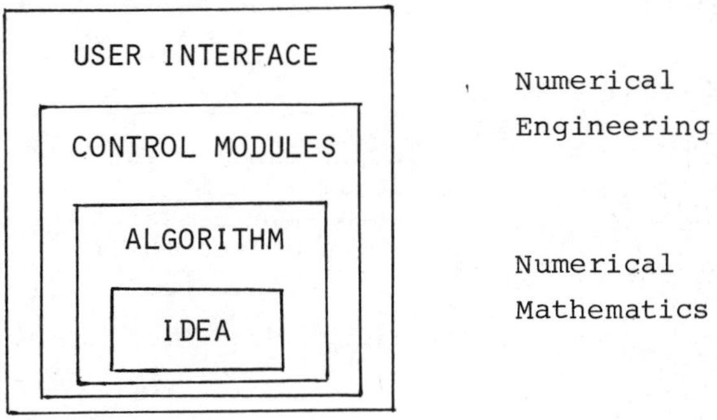

The design of a piece of numerical software normally proceeds in the stages indicated in fig. 1. The generation of the basic approach and its realization in the form of a mathematical algorithm are activities which belong to Mathematics in its classical sense; results in this area may be described in traditional forms, through the proof of assertions under clearly stated hypotheses. The aggregation of these results forms the body of Numerical Mathematics; they are found in the well-known journals devoted to this branch of Mathematics.

However, the *implementation* of any non-trivial numerical algorithm, i.e. its transformation into a program for a digital computer, requires not only the translation of the algorithm proper into a programming language but also the design of suppelementary program modules which guide and control the execution of the algorithm; we will call these *"control modules"*. Finally, the incorporation of the program into a computer library as a piece of numerical software necessitates the addition of communication modules which conform with local conventions and which may also depend on the group of users for which the software is intended.

In the following we will be exclusively concerned with the control layer of numerical software products. The design and the evaluation of the control modules is essentially an engineering activity: These program sections have to achieve certain design objectives some of which may be only intuitively defined, and they must do this for a wide variety of problem data. Reliability and robustness become important design criteria. Furthermore, the computational effort spent within these modules must be kept to a minimum.

Often the adequacy of the design of the control modules decides about the quality of the whole software product: Different implementations of the same numerical algorithm may behave quite differently, particularly in situations near the boundary of the scope of the algorithm.

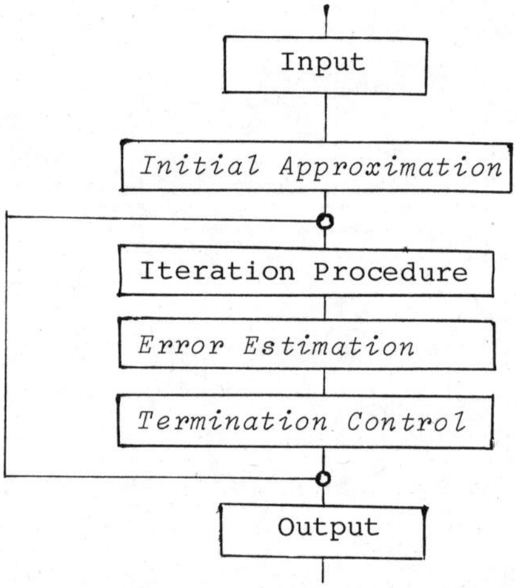

Fig. 2

E.g. in an iterative computation, the iteration procedure would constitute the mathematical core module while the remaining program sections shown in fig. 2 would be considered control modules from our point of view. In their design one obviously has to combine a thorough analysis of the numerical situation with heuristic considerations; some decisions may have to be based on appropriate experimentation.

The claim that such Numerical Engineering constitutes a necessary and important complement of Numerical Mathematics is becoming more and more accepted. However, while other fields of engineering have their established ways of research, of reporting the results of such research, and of judging the merits of such results, this is not the case in the field of designing numerical software. In the following, we will present a few thoughts and suggestions towards an improvment of this situation.

2. MODULAR ANALYSIS AS A BASIS FOR RATIONAL REPORTING ABOUT THE DESIGN OF CONTROL MODULES

What do we expect from a scientific publication about a piece of numerical software which implements a mathematical algorithm which is essentially known? (This may be assumed as the standard situation because a significantly new numerical idea will usually be published separately prior to its implementation.) A typical example is a code for initial value problems in ordinary differential equations based on some pair of embedded Runge-Kutta procedures.

In this case, the primary interest will lie in the design of the control structures of the implementation since they will determine the efficiency, reliability, and robustness of the code. While the details of the realization of these control modules can be read from a well-commented listing of the program, a publication should explain the rationale behind the chosen strategy and the various choices which have been made. As everywhere in science it is the "why" rather than the "how" which represents the essential information.

Thus, a report about some control module should begin with a clear specification of the *design objective(s):* What is this module to achieve and in which situations is it to function properly? Without such a statement of the design goal it will be impossible to judge the quality of the design; the designer may have had a principal objective in mind which is different from what we assume.

Next, the author should explain his *design strategy:* What approach has he chosen to solve his design problem and why? Why not some alternate approach? Normally, a rational presentation of the design principle will not be possible without reference to a mathematical *model* of the computation which is to be controlled. Often, such a model also describes the scope of the design strategy in a natural fashion. Furthermore, it may permit an objective interpretation of the effect of certain numerical parameters which occur in control strategies, like threshold values and the like.

Nevertheless, a good deal of the argumentation in favor of a certain design will retain a subjective element. This is quite natural in an engineering environment and should not be considered "unscientific". However it makes it necessary that it be finally *demonstrated* that the design objective has been achieved. In some sufficiently simple situations it will be possible to do this analytically, within the framework of the model of the computation. More often the demonstration will have to be of an experimental nature. (Again both cases are commonplace in other fields of engineering.)

The experimental evidence of the proper functioning of a given control module within a software product cannot (normally) be achieved by ordinary testing of the complete code. It might easily happen that two rather different realizations of the control module under scrutiny lead to indistinguishable test results only because a poor design in another part of the software disguises the existing differences or because our battery of test problems is not adequate for exhibiting these differences. What we would rather like to see is that the input-output (or cause-effect) relationship which the module is supposed to realize actually occurs, in a wide variety of situations.

For this purpose one should artificially generate the input to our module and observe the output. In a recursive computation, this output will naturally determine the input in the next cycle. We should therefore attempt to devise a *simulation* of the computational process in which only the control module under observation is real while the remaining computation is simplified such that only those aspects are retained which influence the operation of our module, i.e. the correct feedback behavior in a given problem data situation; cf. fig. 3.

Fig. 3

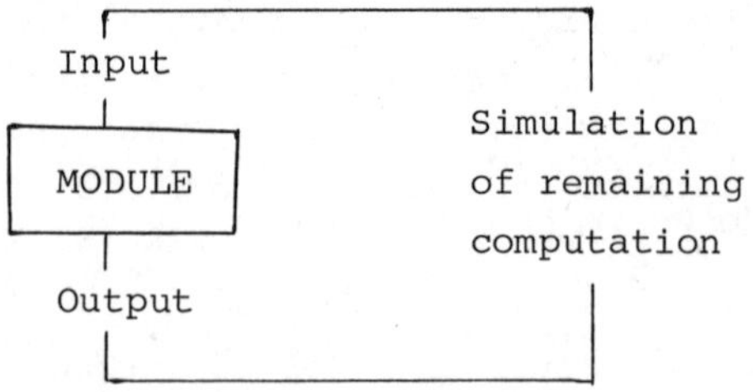

The essential feature of this type of numerical experimentation is the rigid control over the input to the module. Not only is it easy to monitor the input-output behavior of the piece of code under scrutiny but one can freely create input situations which challenge the capabilities of the design of the module in well-defined ways. At the same time, effects of minor design changes or of the choice of threshold parameters can be immediately demonstrated.

The report about the results of a carefully planned and executed set of such simulation experiments will constitute a "proof" of the achievement of the design objectives of a control module, at least within the domain of applications represented by the input data and the simulation model. It will give more authority to an assertion about the potential of that module than could ever be achieved by test runs involving the complete software product, even with suitable monitoring.

This claim is not intended to belittle the importance of testing the complete code. On the contrary, the final "proof" of the functioning of the *whole* program can - at the present state of the art - only be obtained by extensive and well-documented testing. While this will expose the superiority of one code over another, it will, however, rarely give much insight into the reasons for this superiority, in particular into the merits or shortcomings of various individual parts of these codes.

Although we have had the reporting about the design of control modules in mind, we have obviously assumed that the various phases of consideration and experimentation have actually been executed during the design process. In this sense, the establishment of appropriate publication traditions will have a beneficial impact on the quality of numerical software design, and vice versa.

3. A CASE STUDY

For most of the remainder of this lecture we will discuss the design of a particular piece of code: the stepsize control module in an initial value problem solver for (non-stiff) systems of ordinary differential equations. This discussion is not meant as a significant contribution to the state of the art in this field; it is rather to illustrate, in the sense of the last paragraph of section 2, some of the points which have been put forward so far. For the sake of simplicity we will restrict ourselves to a "fixed method" code.

What is the design objective of the stepsize control module? Immediately (and independently of any "philosophical" issues about error control in the numerical solution of o.d.e.) we find that we have three completely different basic objectives depending on the "state" of our computation:

State	Design objective
Starting	Permit transition to ordinary state as "quickly" as possible
Ordinary	Keep an estimate of the "local error" below (but close) to a user-specified accuracy level
Emergency	Permit an "efficient" passage of the difficulty (if possible).

Before we enter into details, let us observe once more that traditional testing does not provide the proper means for an evaluation of the relative merits of alternative designs. If we repeatedly run our code, with variations in the stepsize control module, against the usual battery of test problems and obtain the usual "statistical" evaluation of the results (e.g. via the Toronto Testing Package [1]), we may be able to eliminate a particularly poor design but we cannot hope to recognize more subtle properties of a certain strategy. First of all, we cannot trace the effects to their real source; more important, we may not even realize a crucial sensitivity of a design to a certain computational situation: This situation may not occur within our test problem battery (for this code), or if it does, we will probably not recognize it.

Of course, we may monitor the input and the output of the stepsize control module during these test runs and store the traces from these monitors. I have done that on several such occasions and it has provided some insight; but it leads to piles of diagnostic output which is often difficult to evaluate. Besides, one still has no true control over the input situations which the stepsize control module encounters.

How could we instead evaluate the functioning of the stepsize control module by simulation experiments? We have to generate a sequence of input data to the module, with the proper feedback from the output of the module (cf. fig. 3). In the present case our input consists of

- the current stepsize h_{old} (and previous stepsizes in a multistep code),
- the estimate of the local error L (whatever this is),
- a user-specified accuracy tolerance TOL,
- the current state of the computation.

The output of the stepsize control module is

- the stepsize h_{new} for the next (or repeated) step
- the state of the computation for the next step

If we neglect eventual further input or output data and if we assume that the state is not changed outside our module and that TOL remains fixed throughout, the simulation mechanism has only to create a new value L of the local error, shift h_{new} to h_{old} (and shift the previous stepsizes in a multistep code), and feed these back into the stepsize control module.

How should we generate L as a function of h_{new}? In a code with a *one-step*, RK-like integration procedure the actual computation proceeds thus:

$$\eta_n := \eta_{n-1} + h_n \sum_i b_i \phi_{ni},$$

$$\bar{\eta}_n := \eta_{n-1} + h_n \sum_i \bar{b}_i \phi_{ni},$$

(3.1) $$L := \bar{\eta}_n - \eta_n = h_n \sum_i \lambda_i \phi_{ni},$$

where the ϕ_{ni} are f-evaluations at certain auxiliary η-values. In a standard p-th order procedure, the true local errors of η_n and $\bar{\eta}_n$ are $O(h_n^{p+1})$ and $O(h_n^p)$ resp., L is an approximation of the local error of $\bar{\eta}_n$ (this approach is often called "local extrapolation"). Thus the sum in (3.1) must be a (p-1)st order difference of f-values evaluated near the solution trajectory, or approximately a (p-1)st order difference of values of the derivative of the solution:

(3.2) $$L \approx h_n \sum_i \lambda_i y'(t_{n-1} + c_i h_n)$$

which implies

(3.3) $$L \approx h_n^p \varphi(t_{n-1} + h_n/2) \quad \text{in a smooth situation.}$$

Let us now assume that our stepsize control module (as well as the acceptance module) looks only at some weighted norm of the vector L, as is normally the case. Then (3.3) turns into the *scalar* relation

(3.4) $$L \approx h_n^p \bar{\varphi}(t_{n-1} + h_n/2).$$

It is easily confirmed that, in a smooth situation, (3.4) represents the error estimate generated by the code correctly to within a factor $(1 + O(h_n))$. Thus, a smooth "test problem" of arbitrary dimension affects the stepsize control unit *only* through a non-negative continuous scalar function $\bar{\varphi}$, and we may simulate all possible smooth applications of our code by considering all such functions $\bar{\varphi}$.

In this sense, we may use (3.4) as our *model* of the computation for the design of the stepsize control, at least in the ordinary state. Thus, our attempt at constructing a valid simulation has lead us to an improved insight into what the module is actually up against.

The standard *design strategy* is to aim at $L \approx r\,TOL$, where $r < 1$ is a design parameter. With the model (3.4), this leads to the well-known control relation

$$(3.5) \qquad h_{new} := \left(r\,\frac{TOL}{L}\right)^{1/p} h_{old},$$

which implies, for a continuously differentiable $\bar{\varphi}$,

$$(3.6) \qquad L_n/TOL = \left(1 + h_n \frac{\bar{\varphi}'(t_n)}{\bar{\varphi}(t_n)} + O(h_n^2)\right) r$$

for the sequence $\{L_n\}$ of *norms* of error estimates arising from (3.5) and (3.4) with a given $\bar{\varphi}$. Equation (3.6) which is even correct if we attach the factor $(1 + O(h_n))$ to (3.4) may be used as a suitable basis for an evaluation of the reactions of our control (3.5).

The quantity r in (3.5) can be used for "tuning": We wish to have h_{new} as large as possible without too high a risk of running into $L > TOL$, which leads to a rejection of the step. (3.6) suggests that r should not be given a fixed value but that it should depend on the local (or an average) value of h since $\bar{\varphi}'/\bar{\varphi}$ is a quantity which depends upon the problem (and the integrator) but not upon h. Thus we may try

$$(3.7) \qquad r = \frac{r_o}{1 + c\,h_{old}}, \quad \text{with, say, } r_o = .9,\ c = 2.5,$$

or, since the average value of h will be proportional to $(TOL)^{1/p}$ and we would prefer not to recompute r in every step,

$$(3.8) \qquad r = \frac{r_o}{1 + c(TOL)^{1/p}}, \quad \text{with similar values of } r_o \text{ and } c.$$

Naturally, the choice of c depends on our expectation of the "normal" size of $\bar{\varphi}'/\bar{\varphi}$ while $r_o < 1$ simply adds some more "safety".

The results of simultations displaying the effects of changes in c and r_o will be reported elsewhere. They establish the desirability of having $c > 0$ in (3.8), i.e. of a dependence of r on TOL, contrary to common practice: All popular codes which use (3.5) throughout or under certain conditions have a constant r; a few examples are:

DGEAR	$\approx (5/6)^{p/2}$	(at local order p)
DVERK	$(.9)^6 \approx .53$	
STEP	$.5$	

In no case, more than a heuristic argument has been presented for the particular choice.

Equ. (3.6) also identifies the two possible sources for rejections:

a) $\bar{\varphi}'(t_n)$ is large (in relation to $\bar{\varphi}(t_n)$ and h_n^{-1})
b) $\bar{\varphi}(t_n)$ is small (in relation to $\bar{\varphi}'(t_n)$ and h_n)
and $\dfrac{\bar{\varphi}'}{\bar{\varphi}} > 0$.

While situation a) is a legitimate reason for requiring a repetition of the step (and possibly a transition to emergency state), situation b) is perfectly harmless: It will occur (cf. fig. 4) after φ of (3.3) has changed sign in a scalar problem or after the dominant component of φ has changed sign in a system. Here a control of type (3.5) will

Fig. 4

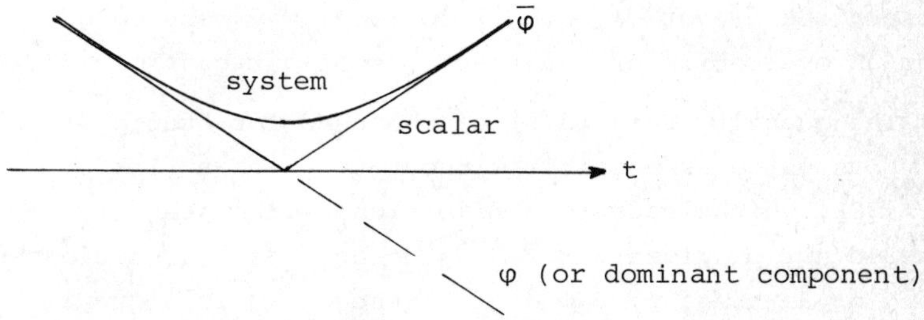

tend to increase the stepsize before the sign change and then find itself with too large steps when φ returns to a normal level. Moreover, since the codes use a fictitious error estimate only, there need be no reason for a stepsize fluctation at all.

To block this silly reaction of their stepsize controls to some extent, all codes introduce a restriction on the admissible increase in stepsize:

(3.9) $$h_{new}/h_{old} \leq q_{max}.$$

This q_{max} is another one of these design parameters which are commonly chosen on purely heuristic grounds; $q_{max} = 2$ is a popular choice. On the other hand, if one introduces the natural objective of keeping the total relative change in h before a sign change of φ below some limit, one finds that q_{max} should depend upon TOL. Again a relation of the kind

$$q_{max} = 1 + d\,(TOL)^{1/p} \tag{3.10}$$

with a value of d to be determined by simulation, seems appropriate and, for small TOL, restricts the undesirable growth of h more effectively than $q_{max} = 2$.

Naturally, such a tight restriction of h_{new}/h_{old} is not suitable during the *starting* state of the computation; its release is one of the means to achieve a transition from a possibly unrealistically small starting stepsize to the natural one within very few steps (cf. our basic design objective for the starting state).

Rather than enter into details of the starting state design we shall turn to an analysis of what we have to expect when we hit a *step discontinuity* in f. Let us assume that the discontinuity occurs at $\hat{t}$ (triggered either through t or y(t)) and define an artificial, smooth continuation $\tilde{y}(t)$ of y(t) and $\tilde{f}(t)$ of f(t,y(t)) beyond $\hat{t}$; cf. fig. 5.

Fig. 5

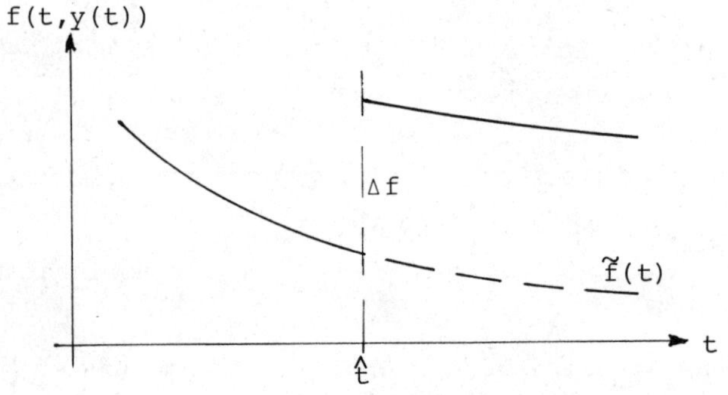

From the original formula (3.1) for L we have for a step which straddles $\hat{t}$

$$\begin{aligned} L &= h_n \sum_i \lambda_i \phi_{ni} \\ &= \underbrace{h_n \sum_i \lambda_i \tilde{\phi}_{ni}}_{\approx h_n^p \tilde{\varphi}(t_{n-1} + \frac{h_n}{2})} + \underbrace{h_n \sum_i{}' \lambda_i \Delta f_{ni}}_{O(h_n) O(\Delta f)} \end{aligned} \tag{3.11}$$

where the second sum includes only those evaluations of f which are past the discontinuity. This sum is no longer a (p-1)-order difference so that it becomes the dominating term; since the norm of the first term will be close to TOL, the norm of the total expression (3.11) will most likely exceed TOL considerably.

To pass $\hat{t}$, the straddling step must be such that $\hat{h}\Delta f$ is of the order of TOL. Obviously, we must strive to discover the position of $\hat{t}$ to within $\hat{h}$ with as few trials as possible; without further knowledge, the natural strategy is "binary search" or bisection.

Fig. 6

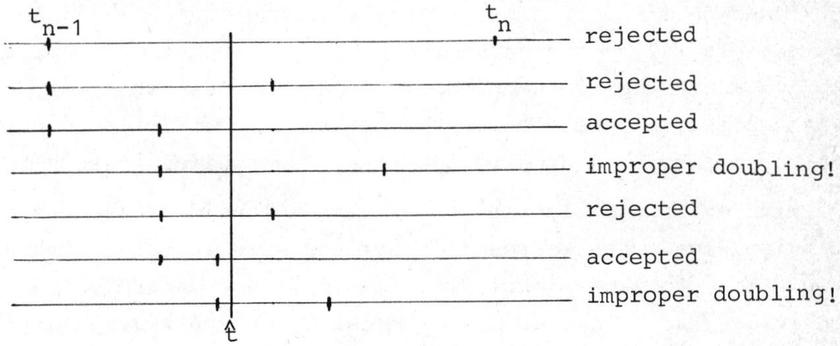

However, if this is done under the ordinary state of stepsize control, with (3.5) and

$$h_{new}/h_{old} \geq q_{min} = \frac{1}{2} \tag{3.12}$$

and $q_{max} = 2$ (cf. (3.9)), each intermediate step which falls short of $\hat{t}$ (and hence has $L \ll TOL$) will be followed by one of doubled length which must fail; cf. fig. 6. This waste is one of the reasons why we should have a special emergency state, with the simple basic control.

$$h_{new} := \begin{matrix} \frac{1}{2}\, h_{old} \\ h_{old} \end{matrix} \quad \text{if the previous step was} \quad \begin{matrix} \text{rejected} \\ \text{accepted.} \end{matrix} \tag{3.13}$$

This will also provide an indicator that we may return from the emergency state: If two successive steps of equal length have been accepted, we must have passed the spot where the "obstacle" has been before; cf. fig. 6.

How do we recognize that we should enter the emergency state? In the case which we have just discussed, (3.11) suggests that we will experience a sudden jump in the size of L. Via (3.5) this will request a sudden drop in h such that h_{new}/h_{old} would be smaller than a suitably chosen threshold q_{min}. Again (3.6) indicates that q_{min} should not be a fixed design parameter (like the 1/2 used in (3.12)) but vary with TOL, perhaps like the inverse of q_{max} of (3.10). The results of simulation runs which have fully confirmed such a strategy will be reported elsewhere.

Finally, when we leave the emergency state we should pass to the starting rather than the ordinary state so that we can quickly recover from the inappropriately short stepsize at which we have arrived during our attempt to cross the singularity.

I hope that I have been able to demonstrate that many decisions in the design of the stepsize control module of a one-step o.d.e. code may be based on a rather rigorous analysis and that the values of design parameters may be chosen on the basis of simulation runs. Thus a publication about the implementation of such a code could very well compete with scientific publications in established fields of engineering - and it could also be judged by the standards applied in such fields.

With *multistep* algorithms, we have a considerably more complex situation, even when we restrict our analysis to codes with a fixed integration procedure (excluding the important variable order codes). There are two principal new aspects: The computational history from a certain number of previous steps is carried along, and costs occur in stepsize changes (recomputation of coefficients or interpolation to new gridpoints).

However, quite a bit of analysis is still possible with respect to stepsize control design after we have found the "model" for the error estimate which takes the place of (3.3) in a smooth situation. It turns out that - for most codes in current use - this is (see, e.g., [2])

$$(3.14) \qquad L \approx P_1(h_n;\text{history})\ \varphi_1(t_n) + P_2(h_n;\text{history})\ \varphi_2(t_n).$$

Here P_1 and P_2 are polynomials in h, of degree p and with a lowest order term h^2 while the functions φ_1 and φ_2 now reflect the local properties of the problem.

Since the structure of the polynomials P_i is known, we may use (3.14) similarly as (3.3) to check the reactions of a given stepsize module and to guide the design of a new one. Also, though with more effort and some further simplification, simulation experiments are possible; they will still provide more insight at less expense than traditional testing of the complete codes.

A few excellent examples of what can be achieved by analysis and modelling in the design and evaluation of a multistep (variable order) code may be found in the well-known book by Shampine and Gordon ([3]) which also furnishes an excellent example of how one can report about the implementation of a sophisticated piece of numerical software.

4. CONCLUSIONS

We have attempted to demonstrate how the design of control modules in numerical software may be based on rigorous analysis to a larger extent. The relative merits of different designs may more reliably be evaluated by simulation experiments than by traditional testing of complete codes. Similarly the values of design parameters within the individual modules may be selected on the basis of such simulations. The detailed presentation of the mathematical models which have guided the design of the various control modules and the description of the simulation (or other) experiments which have established the functioning of the design should be explicitly included in publications about numerical software products. Scientific reporting in other fields of engineering could serve as an example.

It is essential for the further development of Numerical Mathematics that its engineering aspects which are represented by the design, evaluation, and maintenance of numerical software, are fully developed to a level of achievement which is comparable to that of its more theoretical aspects. The establishment of a tradition in reporting about the design of numerical software will give more recognition to the excellent scientific work which is often done in this area, and to those who do it. Also the existence of a collection of well-confirmed or even proven results concerning the design of control modules in numerical software will permit referencing to a much larger extent than it is possible now. It is hoped that the combined efforts of the research scientists working in this area and of the editors and referees of the relevant journals will succeed in establishing such a tradition.

REFERENCES:

[1] W.H. Enright: Using a testing package for the automatic assessment of numerical methods for ODE's, in: Performance Evaluation of Numerical Software, Conference Proceedings, North Holland, 1979.

[2] H.J. Stetter: Interpolation and error estimation in Adams PC-codes, SINUM 16 (1979) 311-323.

[3] L.F. Shampine, M.K. Gordon: Computer solution of ordinary differential equations, the initial value problem, W.H. Freeman, San Francisco, 1975.

THE NUMERICAL SOLUTION OF TURBULENT FLOW PROBLEMS IN GENERAL GEOMETRY

E. L. Wachspress

THE TEACH APPROACH

The Navier-Stokes equations considered here are

$$\nabla \cdot (\rho\vec{v}\vec{v} - \mu\nabla\vec{v}) = -\nabla p + \vec{f} \qquad (1a)$$

$$\nabla \cdot \rho\vec{v} = 0 \qquad (1b)$$

where ρ is the fluid density, μ is its viscosity, $\vec{v} = u\vec{i} + v\vec{j}$ is the velocity, p is pressure, and $f = f_1\vec{i} + f_2\vec{j}$ is a known body force that includes such effects as gravity. Basic TEACH boxes and the TEACH interlocking rectangles are shown in Figure 1.

A TEACH iteration cycle consists of five steps:

1. Equation coefficients that vary during the iteration are computed.
2. The x-momentum equations are used to iterate on u.
3. The y-momentum equations are used to iterate on v.
4. The continuity equation is used to iterate on p.
5. A continuity correction is applied to all u and v components.

The momentum equation is linearized before the velocity iteration by replacing the $\rho\vec{v}\vec{v}$ term by $\rho\vec{v}^*\vec{v}$ with $\vec{v}^*$ the result of Step 5. An input guess is used for $\vec{v}^*$ during the first cycle. The discrete equation associated with unknown u_c in Figure (1a) is derived by integration over the box of the $\vec{i}$-component of Equation (1a) using nodal values in an obvious way:

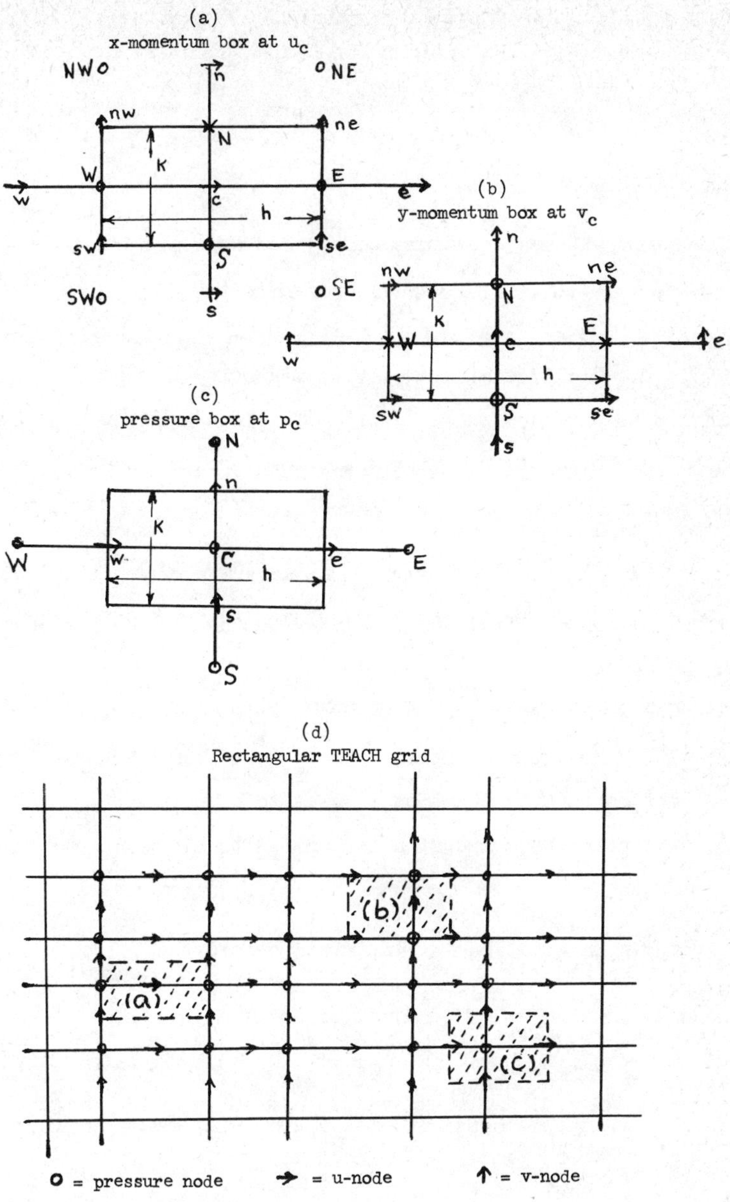

FIGURE 1. TEACH Grid With Interlocking Boxes.

$$k\rho_E u_E^* u_E + h\rho_N v_N^* u_N - k\rho_W u_W^* u_W - h\rho_S v_S^* u_S$$

$$+ \mu_E \frac{k}{h} (u_c - u_e) + \mu_W \frac{k}{h} (u_c - u_w) + \mu_N \frac{h}{k_N} (u_c - u_n) \tag{2}$$

$$+ \mu_S \frac{h}{k_S} (u_c - u_s) = k(p_W - p_E) + hkf_1 .$$

The body force f_1 is evaluated at c; pressures, densities, and viscosities are known at the pressure nodes with results from the last cycle used in the momentum equations. In Equation (2) values at N and S, which are not nodes, are averages of the four neighboring nodal values. For example, $p_N = \frac{1}{4} (p_E + p_W + p_{NE} + p_{NW})$. The starred velocities are also averages of adjacent values. For example, $u_E^* = \frac{1}{2} (u_c^* + u_e^*)$.

The manner in which the unstarred velocity components at pressure nodes are related to the values being computed at velocity nodes is a crucial part of the computation. A form of hybrid differencing due to Spalding[4] is used in TEACH. A Péclet number is defined as follows for the pressure nodes neighboring velocity node c:

$$P_E = \frac{\rho_E u_E^* h}{2\mu_E} \qquad P_W = \frac{\rho_W u_W^* h}{2\mu_W}$$

$$P_N = \frac{\rho_N v_N^* k_N}{2\mu_N} \qquad P_S = \frac{\rho_S v_S^* k_S}{2\mu_S} . \tag{3}$$

If the Péclet number has magnitude less than unity, the unstarred velocity is taken as the average of neighboring values. Thus, if $|P_E| < 1$, then $u_E = \frac{1}{2} (u_c + u_e)$. If the Péclet number is greater than unity in magnitude, then the velocity is taken as the value at its upwind neighbor. Thus, if $P_E > 1$, $u_E = u_c$ and if $P_E < -1$, then $u_E = u_e$.

The net fluid outflow from the box around c,

$$F_c(u) = k(\rho_E u_E - \rho_W u_W) + h(\rho_N v_N - \rho_S v_S) , \tag{4}$$

is zero when mass is conserved. During the course of the iteration, $F_c(u^*)$ is in general nonzero. Stability is enhanced by adding $F_c(u^*)u_c$ on the righthand side of (2) when $F_c(u^*)$ is positive and by adding $|F_c(u^*)|u_c$ on the lefthand side of (2) when $F_c(u^*)$ is negative. Hybrid differencing together with this F_c modification of the difference equations assures a momentum equation coefficient matrix that is irreducibly diagonally dominant and essentially nonpositive. Such properties play a significant role in convergence analysis.

Velocity boundary conditions are introduced as follows:

1. The inlet velocity (on column $i = 1$) is specified.

2. Velocity on a wall (row $j = 1$ and/or $j = J$) is zero.

3. The velocity component normal to a symmetry axis is zero.

4. The exit column ($i = I$) is far enough downstream so that an asymptotic profile may be assumed at exit.

5. The flow profile is normalized at exit to yield exact mass conservation (flow out I = flow in at $i = 1$).

This last condition is achieved by holding the exit velocity fixed during the velocity iteration (Steps 2 and 3 of the iteration cycle) and updating the exit profile just prior to Step 4 by setting $u_{Ij} = u_{I-2j} + d$ with the additive normalization d chosen to give the desired mass balance. This balance is essential for a meaningful pressure computation.

Pressure equations are derived from mass conservation identities. The net fluid outflow from the box around pressure node C in Figure 1c is

$$F_C(\vec{v}) \equiv k_C(\rho_e u_e - \rho_w u_w) + h_C(\rho_n v_n - \rho_s v_s) , \tag{5}$$

where the densities are taken as the averages of values at neighboring pressure nodes. On convergence, $F_C = 0$. The momentum equation at each node where u is computed may be written as

$$u_c = \frac{k_c}{d_c}(p_W - p_E)_c + \frac{h_c k_c f1}{d_c} + (\text{terms in } u_{n,s,e,w}) \ . \qquad (6.1)$$

Similarly, the momentum equation at a v-node is

$$v_c = \frac{h_c}{d_c}(p_S - p_N)_c + \frac{h_c k_c f2}{d_c} + (\text{terms in } v_{n,s,e,w}) \ . \qquad (6.2)$$

Substituting Equations (6.1) and (6.2) into (5), one obtains

$$\rho_e k_C \frac{k_e}{d_e}(p_C - p_E) - \rho_w k_C \frac{k_w}{d_w}(p_W - p_C) + \rho_n h_C \frac{h_n}{d_n}(p_C - p_N)$$

$$- \rho_s h_C \frac{k_s}{d_s}(p_S - p_C) = F_C(\vec{v}) + (\text{terms in } \vec{v}_{n,s,e,w} \text{ and } f) \ . \qquad (7)$$

Pressures p^* used for the velocity iteration yield velocities $\vec{v}^*$ that satisfy Equation (7) but not necessarily the mass conservation condition of $F_C = 0$. A pressure correction p' is defined by $p' = p - p^*$ and a velocity correction $\vec{v}'$ is defined by $\vec{v}' = \vec{v} - \vec{v}^*$. Then, since $F_C(\vec{v}) = 0$, Equation (7) yields the following Poisson-type pressure equation:

$$\rho_e k_C \frac{k_e}{d_e}(p'_C - p'_E) + \rho_w k_C \frac{k_w}{d_w}(p'_C - p'_W) + \rho_n h_C \frac{h_n}{d_n}(p'_C - p'_N)$$

$$+ \rho_s h_C \frac{h_s}{d_s}(p'_C - p'_S) = - F_C(\vec{v}^*) + (\text{terms in } \vec{v}'_{n,s,e,w}) \ . \qquad (8)$$

Entrance and exit velocities are not computed from momentum equations so that p_W does not appear in pressure equations on Column i = 2 and p_E does not appear in pressure equations on Column I - 1. There are no pressure nodes on Columns 1 and I, there being only u-nodes on these columns. Thus, boundary conditions are incorporated in Equation (8) in a natural way. The last term on the righthand side of Equation (8), involving the corrections to the neighboring velocities, is assumed to be negligible. This leads to error in the computed p' values and the need for iteration cycles. The $F_C(\vec{v}^*)$ term is computed from Equation (5), and the pressure equations may be expressed in matrix notation as $P\,\underline{p}' = -\underline{f}$, in which the coefficient matrix is symmetric and singular. The row sums of P are zero. Only pressure differences appear in Equation (8) and there is an arbitrary additive normalization which is fixed by prescribing the pressure at a "reference" node. The pressure equations can have a solution only when $\underline{f}$ is orthogonal to the null space of P. The sum of the components of $\underline{f}$ must vanish. The normalization of exit flow now gains motivation, for the sum of the components of $\underline{f}$ is equal to the flow out Column I minus the flow in Column i = 1. This has been forced to zero after each Step-3 of a cycle.

The pressure-correction equations are solved by successive over-relaxation with $\underline{p}'$ allowed to float. After this iteration, the values are normalized to maintain the pressure at the reference node:

$$\underline{p}'_{\text{normalized}} = \underline{p}' - \underline{p}'_{\text{reference node}} \tag{9.1}$$

$$\underline{p}_{\text{updated}} = \underline{p}^* + \underline{p}'_{\text{normalized}} . \tag{9.2}$$

The velocity correction for Step 5 of the cycle is

$$u_{c\ \text{updated}} = u_c^* + \frac{k_c}{d_c}(p'_W - p'_E)_c \ ; \tag{10.1}$$

$$v_{c\ \text{updated}} = v_c^* + \frac{h_c}{d_c}(p'_S - p'_N)_c \ . \tag{10.2}$$

The nonlinearity of Equation (1) complicates convergence analysis. Meaningful results are obtained by analysis of linearized equations. Such analysis considers the iteration cycle without Step 1, and the need for iteration cycles then arises from the dropping of the last term in

Equation (8). The scheme thus far described is divergent. A velocity underrelaxation parameter a_V in the interval (0,1) is introduced as follows: $d_C u_C / a_V$ replaces $d_C u_C$ and $(1/a_V - 1) d_C u_C^*$ is added on the righthand side of the momentum equation at each u-node; $d_C v_C / a_V$ replaces $d_C v_C$ and $(1/a_V - 1) d_C v_C^*$ is added on the righthand side of the momentum equation at each v-node. As a_V approaches unity, the underrelaxation disappears but the scheme becomes divergent. As a_V approaches zero, convergence worsens in that the updated velocity approaches the old value. The choice of a_V is critical in a TEACH computation. Convergence characteristics with and without Step 1 are not greatly different.

THE TURF GENERALIZATION

Geometry

The isoparametric parabola through three ordered points with coordinate vectors $\underline{r}_i$ (i = 1, 2, 3) has the parametrization:

$$\underline{r} = \tfrac{1}{2}q(1 + q)\underline{r}_3 - \tfrac{1}{2}q(1 - q)\underline{r}_1 + (1 - q^2)\underline{r}_2 \ .$$

Eight ordered points determine a four-sided isoparametric box with each side an isoparametric parabolic arc. TEACH rectangles are replaced by isoparametric boxes in TURF. Nonorthogonality of grid lines leads to significant complications. It becomes necessary to compute both components of velocity at velocity nodes and to introduce another set of pressure nodes.

A local coordinate system is defined at each pressure node in terms of the coordinates of Pressure Node C and the four neighboring velocity nodes:

$$\underline{a}^T = (a_0, a_1, \cdot\cdot\cdot, a_4) = \left(x_c, \frac{x_e - x_w}{2}, \frac{x_n - x_s}{2}, \frac{x_e + x_w - 2x_c}{2}, \frac{x_n + x_s - 2x_c}{2} \right);$$

$$\underline{b}^T = (b_0, b_1, \cdot\cdot\cdot, b_4) = \left(y_c, \cdot\cdot\cdot, \frac{y_n + y_s - 2y_c}{2} \right);$$

$$\begin{bmatrix} x \\ y \end{bmatrix} = \begin{bmatrix} \underline{a}^T \\ \underline{b}^T \end{bmatrix} \begin{bmatrix} 1 \\ \xi \\ \eta \\ \xi^2 \\ \eta^2 \end{bmatrix} . \tag{11}$$

On Column i = 1, $(a_3, b_3) = (0,0)$ and $(a_1, b_1) = (x_e - x_C, y_e - y_C)$, and on Column i = I, $(a_3, b_3) = (0,0)$ and $(a_1, b_1) = (x_C - x_w, y_C - y_w)$. On Row j = 1, $(a_4, b_4) = (0,0)$ and $(a_3, b_3) = (x_n - x_C, y_n - y_C)$, and on Row j = J, $(a_4, b_4) = (0,0)$ and $(a_3, b_3) = (x_C - x_s, y_C - y_s)$.

In the notation of Figure 2, base covariant and contravariant vectors, the transformation Jacobian, and some useful vector identifies are:

$$\vec{g}_1 = x_\xi \vec{i} + y_\xi \vec{j},\ \vec{g}_2 = x_\eta i + y_\eta j,\ |J| = (x_\xi y_\eta - x_\eta y_\xi) > 0$$

$$\vec{g}^1 = \frac{1}{|J|}(y_\eta i - x_\eta j),\ \vec{g}^2 = \frac{1}{|J|}(-y_\xi \vec{i} + x_\xi \vec{j})$$

$$\vec{g}_m \cdot \vec{g}^n = \delta_{mn} = \begin{cases} 1, & m = n \\ 0, & m \neq n \end{cases} \qquad J\left(\frac{x,y}{\xi,\eta}\right) = \begin{bmatrix} x_\xi & x_\eta \\ y_\xi & y_\eta \end{bmatrix}$$

$$J\left(\frac{\xi,\eta}{x,y}\right) = J^{-1}\left(\frac{x,y}{\xi,\eta}\right) = \frac{1}{|J|}\begin{bmatrix} y_\eta & -x_\eta \\ -y_\xi & x_\xi \end{bmatrix} = \begin{bmatrix} \xi_x & \xi_y \\ \eta_x & \eta_y \end{bmatrix} \tag{12}$$

$$\nabla f = f_\xi \vec{g}^1 + f_\eta \vec{g}^2 \qquad \vec{f} = f^1 \vec{g}_1 + f^2 \vec{g}_2$$

$$\nabla\cdot\vec{f} = \nabla f^1 \cdot \vec{g}_1 + f^1 \nabla\cdot\vec{g}_1 + \nabla f^2 \cdot \vec{g}_2 + f^2 \nabla\cdot\vec{g}_2$$

$$= f^1_\xi + \frac{f^1 |J|_\xi}{|J|} + f^2_\eta + \frac{f^2 |J|_\eta}{|J|} .$$

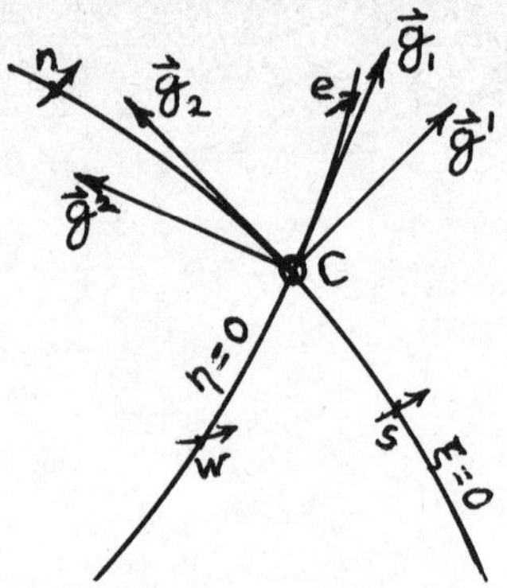

FIGURE 2. Coordinate System at C.

A section of a TURF grid is shown in Figure 3. The vector momentum equation is integrated over each velocity-node box to yield two difference equations associated with the two velocity components at the center node. The mass balance over each pressure-centered box yields the pressure-correction equation associated with the central pressure node.

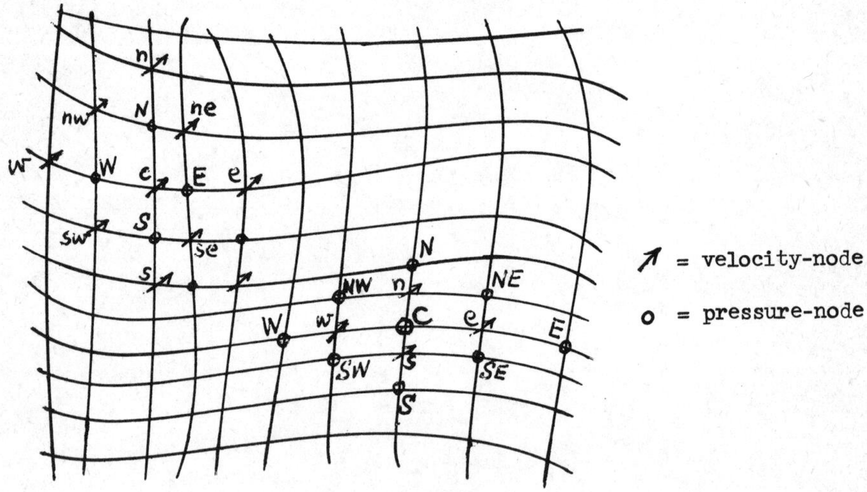

FIGURE 3. TURF Grid.

Momentum Difference Equations

There are three terms in the momentum equation: the transport term $\nabla \cdot \rho\vec{v}^*\vec{v}$, the viscous diffusion-type term $\nabla \cdot \mu\nabla\vec{v}$, and the pressure term ∇p. The integral over the box of the momentum equation is reduced to an integration over the boundary by Gauss's divergence theorem. Contributions of each of the three terms are approximated as follows:

$$\iint_c \nabla \cdot \rho\vec{v}^*\vec{v}dvol = \int_{\partial c} n \cdot \rho\vec{v}^*\vec{v}d\ell = T_E + T_N + T_W + T_S \qquad (13)$$

where T_E is the integral over the east boundary between se and ne, etc.

$$\begin{aligned} T_E &= \int_{-1}^{+1} d\eta |J| \vec{g}^1 \cdot \rho\vec{v}^*\vec{v} \approx \rho_E \int_{-1}^{1} d\eta (y_\eta u_E^* - x_\eta v_E^*)\vec{v}_E \\ &= \rho_E \left[(y_{ne} - y_{se})u_E^* - (x_{ne} - x_{se})v_E^*\right] \vec{v}_E \\ &= 2\rho_E(b_{2E}u_E^* - a_{2E}v_E^*)\vec{v}_E\ , \end{aligned} \qquad (14)$$

and similar expressions are obtained for T_N, T_W, and T_S.

$$-\iint_c \nabla \cdot \mu\nabla\vec{v}dvol = -\int \vec{n} \cdot \mu\nabla\vec{v}d\ell = V_E + V_N + V_W + V_S$$

with $V_E = -\int_{se}^{ne} \vec{n} \cdot \mu\nabla\vec{v}d\ell \approx -\mu_E \int_{-1}^{1} (y_\eta \vec{v}_{x,E} - x_\eta \vec{v}_{yE})d\eta.$

The velocity derivatives are approximated by

$$\vec{v}_{x,E} = \vec{v}_{\xi,E}\xi_{x,E} + \vec{v}_{\eta,E}\eta_{x,E} = \frac{1}{|J_E|}\left(b_{2E}\frac{\vec{v}_e - \vec{v}_c}{2} - b_{1E}\frac{\vec{v}_{ne} - \vec{v}_{se}}{2}\right)$$

$$\vec{v}_{y,E} = \vec{v}_{\xi,E}\xi_{y,E} + \vec{v}_{\eta,E}\eta_{y,E} = \frac{1}{|J_E|}\left(- a_{2E}\frac{\vec{v}_e - \vec{v}_c}{2} + a_{1E}\frac{\vec{v}_{ne} - \vec{v}_{se}}{2}\right),$$

and the viscous contribution from the east boundary is

$$T_E = \frac{\mu_E}{(a_1b_2 - a_2b_1)_E}\left[(a_{2E}^2 + b_{2E}^2)(\vec{v}_c - \vec{v}_e) + (a_{1E}a_{2E} + b_{1E}b_{2E})(\vec{v}_{ne} - \vec{v}_{se})\right].$$

The north, west and south contributions are obtained similarly. Note that the last term in the above equation drops out when the lines connecting the east and west velocity nodes and the north and south nodes are orthogonal. The Péclet number is determined by the ratio of the transport to viscous coefficients. Thus,

$$P_E = \frac{\rho_E(b_{2E}u_E^* - a_{2E}v_E^*)}{\mu_E \dfrac{(a_{2E}^2 + b_{2E}^2)}{(a_{1E}b_{2E} - a_{2E}b_{1E})}}. \tag{15}$$

The pressure term in the momentum equation is:

$$-\iint_c \nabla p \, dvol = -\int \vec{n} p \, d\ell = R_E + R_N + R_W + R_S \; ;$$

$$R_E = -p_E \int_{se}^{ne} \vec{n} \, d\ell = -2p_E(b_{2E}\vec{i} - a_{2E}\vec{j}) . \tag{16}$$

R_N, R_W, and R_S are obtained similarly.

Boundary conditions must be handled with great care and will not be considered here. The purpose of this discussion is to demonstrate how the local coordinates facilitate discretization.

Pressure Correction Equations

The mass balance around pressure box C is

$$F_C(\vec{v}) = F_{Ce} - F_{Cw} + F_{Cn} - F_{Cs}, \text{ where } F_{Ce} = \rho_e\Big[(b_{oNE} - b_{oSE})u_e - (a_{oNE} - a_{oSE})v_e\Big],$$

and the other terms are computed similarly. The vector momentum equation at e may be expressed in the form

$$D_e \underline{v}_e = P_e \underline{p}_e + [\text{other terms as in Equation (6)}]$$

with $\underline{v}_e^T = (u, v)_e$ and $\underline{p}_e^T = (p_E \text{ of } e, p_N \text{ of } e, p_W \text{ of } e, p_S \text{ of } e)$.

Matrix D is a 2 x 2 nonsingular matrix and P is of order 2 x 4. Thus, $\underline{v}_e = D_e^{-1} P_e \underline{p}_e$ + (other terms) and the contribution to F_{Ce} with the "other terms" neglected is

$$F_{Ce} = \rho_e \left[(b_{oNE} - b_{oSE}), (a_{oSE} - a_{oNE}) \right] D_e^{-1} P_e \underline{p}_e . \qquad (17)$$

Pressure equations analogous to Equation (8) are generated by the technique already described for TEACH.

TURF Iteration Cycle

The TURF difference equations have a more complicated structure than the TEACH equations. The vector momentum equation at velocity node c couples the velocity vector at c to the velocity vectors at its e, n, w, and s neighbors and also to its ne, nw, sw, and se neighbors. The latter coupling is, in general, weaker and vanishes when the grid lines are chosen to be orthogonal. Stability analysis suggests an iteration in which first velocities on odd columns are improved with a few grid sweeps and then velocities on even columns are iterated. Two or three such odd-even iterations constitute one velocity update. Difference-equation coupling between odd and even columns is relatively weak and arises only from the nonorthogonality of grid lines. Numerical studies have verified the adequacy of this odd-even iteration. The same procedure is followed with the pressure-correction equations. The odd-even inner iterations increase computation time so that TURF is more demanding of computer time than TEACH. Otherwise, the computation strategy is quite similar for the two programs.

Boundary conditions on velocities and pressures are interrelated. Momentum equations at boundaries are adjusted so that mass balances around pressure boxes yield pressure-correction equations that retain the basic TEACH properties. The pressure coefficient matrix has zero row-sums and is thus singular. The sums of the components of the forcing vectors for both odd and even pressure equations vanish. Convergence and stability analysis for TURF is more complicated than for TEACH, but numerical studies support the adequacy of the geometry generalization.

TEACH CONVERGENCE ANALYSIS

Let $\underline{v}$ be a vector with nodal values of u and v as its components and let $\underline{p}$ be a vector with nodal values of pressure as its components. Let the flow be incompressible with constant fluid density. Then the linearized Navier-Stokes equations may be written as

$$\begin{aligned} (D - L)\underline{v} + B\underline{p} &= \underline{f} \\ C\underline{v} &= \underline{g} \ , \end{aligned} \tag{18}$$

where D is a diagonal matrix with positive diagonal elements and L is a nonnegative matrix with zero entries on its diagonal. Diagonal dominance of D - L yields a spectral radius less than unity for $D^{-1}L$. Let the generalized inverse of X be denoted by X^+. Except for an arbitrary additive normalization of the pressure vector, the unique solution to (18) is

$$\begin{aligned} \underline{p} &= [C(D - L)^{-1}B]^{+}[C(D - L)^{-1}\underline{f} - \underline{g}] \\ \underline{v} &= (D - L)^{-1}(\underline{f} - B\underline{p}) \ . \end{aligned} \tag{19}$$

Asymptotic convergence analysis deals with the iterative solution of Equation (18) by Steps 2-5 of the TEACH cycle repeated in an attempt to obtain the solution in Equation (19). Rather general fixed-point theorems relate actual convergence to asymptotic convergence. If the initial guess is sufficiently close to the solution, then asymptotic convergence is sufficient. In any case, asymptotic convergence is a necessary property.

The TEACH iteration cycle is described by the equations:

$$
\begin{aligned}
\underline{p}_t &= \underline{p}_{t-1} + \frac{1}{a}(CD^{-1}B)^{+}(C\underline{v}_{t-1} - \underline{g}) \\
\underline{v}_t^{*} &= \underline{v}_{t-1} - aD^{-1}B(\underline{p}_t - \underline{p}_{t-1}) \qquad (20) \\
\left(\frac{1}{a}D - L\right)\underline{v}_t + B\underline{p}_t &= \left(\frac{1}{a} - 1\right)D\underline{v}_t^{*} + \underline{f} ,
\end{aligned}
$$

which may be expressed as the matrix equation:

$$
\begin{bmatrix} I & 0 \\ (2-a)B & \left(\frac{1}{a}D-L\right) \end{bmatrix}
\begin{bmatrix} \underline{p}_t \\ \underline{v}_t \end{bmatrix}
=
\begin{bmatrix} I & \frac{1}{a}(CD^{-1}B)^{+}C \\ (1-a)B & \left(\frac{1}{a}-1\right)D \end{bmatrix}
\begin{bmatrix} \underline{p}_{t-1} \\ \underline{v}_{t-1} \end{bmatrix}
+
\begin{bmatrix} -\frac{1}{a}(CD^{-1}B)^{+}\underline{g} \\ \underline{f} \end{bmatrix}
\qquad (21)
$$

The error vector $\underline{e}_t = \begin{bmatrix} \underline{e}_1 \\ \underline{e}_2 \end{bmatrix} = \begin{bmatrix} \underline{p}_t - \underline{p} \\ \underline{v}_t - \underline{v} \end{bmatrix}$ satisfies $\underline{e}_t = T\underline{e}_{t-1}$, where

$$
T = \begin{bmatrix} I & \frac{1}{a}(CD^{-1}B)^{+}C \\ -a(I-aD^{-1}L)^{-1}D^{-1}B & (I-aD^{-1}L)^{-1}[(1-a)I - (2-a)D^{-1}B(CD^{-1}B)^{+}C] \end{bmatrix} .
$$

The TEACH iteration is asymptotically convergent when the spectral radius* of T is less than unity. Matrices X and G are defined as

$$
X = (I - aD^{-\frac{1}{2}}LD^{-\frac{1}{2}})^{-1} \text{ and } G = D^{-\frac{1}{2}}B(CD^{-1}B)^{+}CD^{-\frac{1}{2}} . \qquad (22)
$$

Matrix G is a projection matrix ($G^2 = G$) and, hence, has eigenvalues of unity and zero only. In TEACH, $B = C^T$ and G is symmetric. The range and null space of G are orthogonal in TEACH. This is not the case in TURF. However, the theory generalizes to TURF when one allows for nonorthogonality.

---

*Let $\underline{1}$ denote the vector whose components are all unity. Then $B\underline{1} = \underline{0}$ and $T\underline{e}' = \underline{e}'$ for $\underline{e}_1' = \underline{1}$ and $\underline{e}_2' = \underline{0}$. We exclude this eigensolution from the spectral analysis of T since there is no error associated with $\underline{e}'$. Its component in the numerical solution is set by the additive pressure normalization.

If $\underline{w} = \begin{bmatrix} \underline{w}_1 \\ \underline{w}_2 \end{bmatrix}$ is an eigenvector of T with eigenvalue λ, then

$$\underline{w}_1 + \frac{1}{a}(CD^{-1}B)^{+}C\underline{w}_2 = \lambda \underline{w}_1$$

$$\underline{w}_1 = -\frac{1}{a(1-\lambda)}(CD^{-1}B)^{+}C\underline{w}_2, \text{ and}$$

$$-a(I - aD^{-1}L)^{-1}D^{-1}B\underline{w}_1$$
$$+ (I - aD^{-1}L)^{-1}[(1-a)I - (2-a)D^{-1}B(CD^{-1}B)^{+}C]\underline{w}_2 = \lambda \underline{w}_2 .$$

Eliminating $\underline{w}_1$ and combining terms, one obtains

$$\left\{\left[\frac{1}{1-\lambda} - (2-a)\right](I - aD^{-1}L)^{-1}D^{-1}B(CD^{-1}B)^{+}C + (1-a)(I - aD^{-1}L)^{-1} - \lambda I\right\}\underline{w}_2 = \underline{0} .$$

Multiplying on the left with $D^{\frac{1}{2}}$ and defining $\underline{z} = D^{\frac{1}{2}}\underline{w}$, one obtains

$$\left\{\left[\frac{1}{1-\lambda} - (2-a)\right]XG + (1-a)X - \lambda I\right\}\underline{z} = \underline{0} . \qquad (23)$$

Multiplication of (23) on the left by X^{-1} now yields

$$\left\{\left[\frac{1}{1-\lambda} - (2-a)\right]G + (1-\lambda-a)I + \lambda a D^{-\frac{1}{2}}LD^{-\frac{1}{2}}\right\}\underline{z} = \underline{0} . \qquad (24)$$

The unit vector $\underline{1}$, all of whose elements are unity, is in the null space of G. Let $(\lambda_1, \underline{z}_1)$ be the eigensolution of Equation (24) with eigenvalue closest to unity. Multiplication of Equation (24) with this eigensolution on the left by $\underline{1}^T$ gives:

$$(1-\lambda_1-a)\underline{1}^T\underline{z}_1 + a\lambda_1\underline{1}^T D^{-\frac{1}{2}}LD^{-\frac{1}{2}}\underline{z}_1 = 0 ,$$

and

$$\lambda_1 = \frac{1-a}{1 - a\dfrac{\underline{1}^T D^{-\frac{1}{2}}LD^{-\frac{1}{2}}\underline{z}_1}{\underline{1}^T\underline{z}_1}} . \qquad (25)$$

Let $D^{-\frac{1}{2}}LD^{-\frac{1}{2}}\underline{x}_i = s_i\underline{x}_i$, and let $\underline{x}_1$ be the positive eigenvector with eigenvalue s_1 equal to the spectral radius of $D^{-\frac{1}{2}}LD^{-\frac{1}{2}}$. The existence of this eigenvector follows from the nonnegativity of the matrix. Diagonal dominance of D - L ensures $s_1 < 1$. Eigenvector $\underline{x}_1$ is a good approximation to $\underline{z}_1$ for Equation (25). This yields the estimate:

$$\lambda_1 \approx \frac{1 - a}{1 - as_1} . \qquad (26)$$

Moreover,

$$\frac{\partial\lambda_1}{\partial a} \approx - \frac{1 - s_1}{(1 - as_1)^2} \qquad (27)$$

and it is seen that this eigenvalue estimate decreases from unity when a = 0 to zero when a = 1.

The other end of the spectrum of T must now be examined. As the underrelaxation parameter is increased, an eigenvalue of T moves through -1, and the value a' at which an eigenvalue of -1 occurs is an upper bound on the underrelaxation. Equation (24) yields:

$$\left[\left(a' - \frac{3}{2}\right)G + (2 - a')I\right]\underline{z}_{-1} = a'D^{-\frac{1}{2}}LD^{-\frac{1}{2}}\underline{z}_{-1} \qquad (28)$$

Let $\underline{z}_{-1} = \underline{r} + \underline{n}$ with $\underline{r}$ in the range of G and $\underline{n}$ in the null space of G. Since G is symmetric, it follows that $\underline{r}^T\underline{n} = 0$. Substituting $\underline{r} + \underline{n}$ for $\underline{z}_{-1}$ in Equation (28) and multiplying on the left by $\underline{r}^T + \underline{n}^T$ yields

$$\begin{aligned} \frac{1}{2}|\underline{r}|^2 + (2 - a')|\underline{n}|^2 &= a'(\underline{r}^T + \underline{n}^T)D^{-\frac{1}{2}}LD^{-\frac{1}{2}}(\underline{r} + \underline{n}) \\ &= \frac{1}{2}a'(\underline{r}^T + \underline{n}^T)D^{-\frac{1}{2}}(L + L^T)D^{-\frac{1}{2}}(\underline{r} + \underline{n}) \\ &< a'(|\underline{r}|^2 + |\underline{n}|^2) , \end{aligned}$$

by virtue of the row and column diagonal dominance of D - L.

Therefore,

$$a' > \frac{1}{2} \frac{|\underline{r}|^2 + 4|\underline{n}|^2}{|\underline{r}|^2 + 2|\underline{n}|^2} \geq \frac{1}{2} . \tag{29}$$

There are in fact eigenvectors of $D^{-\frac{1}{2}}LD^{-\frac{1}{2}}$ with eigenvalues close to unity that have small components in the null space of G. The above bound on a' is quite realistic.

It has thus been shown that the range of underrelaxation for which convergence may be expected is $(0,\frac{1}{2})$ with the optimum value being close to $\frac{1}{2}$. In each iteration cycle, the velocity and pressure nodal values are updated by iteration. Hence, Equations (20) are not satisfied exactly. Matrix T is the correct matrix for error analysis only when the iterative solution of Equation (20) is well converged. The effect of inner iteration on cycle convergence is not easily predicted. In practice it is sometimes necessary to choose an underrelaxation somewhat less than $\frac{1}{2}$. This convergence analysis suggests that more attention should be directed to the inner iteration in such cases. For high Reynolds-number flow, the velocity iteration converges quite rapidly. The pressure iteration than plays an important role in determining the overall convergence characteristics. The pressure equations are Laplacian in structure and methods for both direct and iterative solution of these equations have been analyzed exhaustively. Very efficient solution techniques are known. Sufficient effort should be spent on the pressure equations so as not to hamper cycle convergence.

In some cases, cycle convergence may be accelerated by extrapolation. However, the nonlinear transport term in the equations limits the benefits of such extrapolation. An additional complication is introduced by turbulence modeling which leads to simultaneous iteration on the viscosity.

Choice of $a = \frac{1}{2}$ assures stability in the sense that the eigenvalues of T are greater than -1. Cycle convergence is governed by the proximity to unity of the largest eigenvalue. If $s_1 = 1 - \epsilon$ with $0 < \epsilon \ll 1$, then Equation (26) yields an estimate of this eigenvalue of

$$\lambda_1 \approx \frac{1 - a}{1 - a(1 - \epsilon)} = \left[1 + \frac{a\epsilon}{1 - a}\right]^{-1} \approx 1 - \frac{a\epsilon}{1 - a} . \tag{30}$$

and when $a = \frac{1}{2}$: $\lambda_1 \approx 1 - \epsilon$. Thus, the spectral radius of $D^{-1}L$ approximates the eigenvalue of T closest to unity. Although it has not been shown that this is the spectral radius of T, numerical experience supports this hypothesis. The theoretical difficulty arises from lack of a proof that T has no complex eigenvalue of greater magnitude.

REFERENCES

1. Gosman, A. D., B. E. Launder, and J. H. Whitelaw. "Turbulent Recirculating Flow-Prediction and Measurement." Penn State University Lecture Notes (July 28 - August 1, 1975).
2. Zienkiewicz, O. C. The Finite Element Method. McGraw Hill, U. K. Ltd. 3rd expanded edition (1977).
3. Wachspress, E. L. Iterative Solution of Elliptic Systems. Prentice Hall, Englewood Cliffs, N. J. (1966).
4. Spalding, D. B. "A Novel Finite-Difference Formulation for Differential Equations Involving Both First and Second Derivatives." Int. J. Num. Methods in Eng'g., 4 (1972).

THE RATE OF CONVERGENCE OF A MULTIPLE GRID METHOD

P. Wesseling

1. INTRODUCTION

The purpose of this paper, which is a polished version of [16], is to study the rate of convergence and the computational complexity of a so-called multiple grid method for the solution of the linear algebraic system that results from the use of a finite difference method for the numerical solution of the following partial differential equation:

$$-(a_{ij}u_{,i})_{,j} - (b_i u)_{,i} + cu = f, \tag{1.1}$$

with a_{ij}, b_i, c, f and u functions of two variables x_1, x_2 with $(x_1,x_2) \in \Omega \subset \mathbb{R}^2$. Cartesian tensor notation is used. It is assumed that the coefficients are sufficientl smooth, and furthermore:

$$a_{ij} = a_{ji}, \quad B_1\xi_i\xi_i \leq a_{ij}\xi_i\xi_j \leq B_2\xi_i\xi_i, \ \forall\ \xi_i \in \mathbb{R}\ , \ B_1 > 0 \tag{1.2}$$

in $\bar{\Omega}$, i.e. (1.1) is uniformly elliptic. The boundary condition is:

$$u\big|_{\partial\Omega} = 0. \tag{1.3}$$

The region Ω is specified to be the unit-square $(0,1) \times (0,1)$.

In order to apply the finite difference method a computational grid Ω^ℓ is defined as follows:

$$\Omega^\ell \equiv \{(x_1,x_2) |\ x_i = m_i . 2^{-\ell},\ m_i = 0(1)2^\ell\},\ \ell \in \mathbb{N}\ . \tag{1.4}$$

A set of grid-functions U is defined as follows:

$$U^\ell \equiv \{u^\ell : \mathbb{Z} \times \mathbb{Z} \to \mathbb{R}\, |\, u^\ell_{ij} = 0 \ \text{ for i and j outside } (0,2^\ell)\}, \tag{1.5}$$

where the subscripts i,j indicate the function value at the grid-point $(i.2^{-\ell},j.2^{-\ell})$. The domain of u^ℓ is extended outside Ω^ℓ in order to facilitate the application of finite difference operators near $\partial\Omega \cap \Omega^\ell$.

The algebraic system of equations to which the multiple grid will be applied is denoted as follows:

$$A^\ell u^\ell = f^\ell, \tag{1.6}$$

with u^ℓ, $f^\ell \in U^\ell$ and A^ℓ a $d_\ell \times d_\ell$ matrix, with $d_\ell = (1+2^\ell)^2$ the number of points of Ω^ℓ.

The multiple grid method makes use of a hierarchy of computational grids Ω^k,

$k = \ell-1,\ell-2,\ldots,$ with d_k points, $d_k < d_{k+1}$ and a corresponding hierarchy of sets of grid-functions U^k; Ω^k and U^k are defined by (1.4) and (1.5) respectively, with ℓ replaced by k. The simultaneous use of computational grids Ω^k is the reason for the appellation "multi-level method" (cf. [5], [6]) or "multiple grid method" (cf. [12]). Because Ω^{k-1} contains fewer points than Ω^k we will call Ω^{k-1} "coarser" than Ω^k. On U^k an inner product and a norm are defined as follows:

$$(u^k,v^k)_k \equiv 4^{-k} \sum_{i,j=-\infty}^{\infty} u^k_{ij}v^k_{ij} \ , \quad ||u^k||_{0,k} \equiv (u^k,u^k)_k^{\frac{1}{2}}. \tag{1.7}$$

Multiple grid methods are applicable to much more general problems, see for example [5,15]. Also, more and more general proofs concerning the rate of convergence are appearing. In the last section a short survey of the literature will be given. Because completely general proofs are rather complicated it is thought useful to present a proof here for the case of a fairly general equation (1.2), but the simple boundary condition (1.3) and simple region (1.4). It will turn out that the asymptotic computational complexity of the method is $O(\ell 4^\ell)$, which is superior to any other method that the author knows of, for this degree of generality.

2. A MULTIPLE GRID ALGORITHM

Let there be defined restriction operators r^k and prolongation operators p^k:

$$r^k : U^k \to U^{k-1}, \ p^k : U^{k-1} \to U^k, \ k = \ell,\ell-1,\ell-2,\ldots \tag{2.1}$$

An example will be given later. We define in quasi-ALGOL:

<u>Algorithm 1</u>

```
k:=ℓ+1;
start:k:=k-1, u^{k,0}:=0;
for μ:=0 step 1 until σ-1 do
begin u^{k,μ+½}:=M^k(u^{k,μ},f^k);
   f^{k-1}:=r^k(f^k-A^k u^{k,μ+½});u^{k,μ+1}:=u^{k,μ+½}+p^k v^{k-1};
   comment For the definition of v^{k-1} see below.
end of algorithm 1;
```

The function v^{k-1} is some approximation of $u^{k-1} \equiv (A^{k-1})^{-1}f^{k-1}$, which satisfies:

$$||v^{k-1}-u^{k-1}||_{0,k} \le \delta_{k-1}||u^{k-1}||_{0,k}. \tag{2.2}$$

The matrix A^{k-1} and the parameter δ_{k-1} are to be chosen later.
The way in which v^{k-1} is to be obtained is not specified. M^k is shorthand for some relaxation algorithm.

Algorithm 1 is a two-grid method. The following algorithm is a true multiple grid method, in which the number of grids is arbitrary:

Algorithm 2 As algorithm 1, with v^{k-1}, $k=\ell(-1)j+2$ computed by algorithm 1 starting at the label start, while v_j is computed exactly by some direct method.

It follows that $v^{k-1} = u^{k-1,\sigma}$, $v^j = u^j$.

For the motivation of algorithms 1 and 2 we refer to [6], where the intuitive background of multiple grid methods is amply elucidated.

3. A CONVERGENCE THEOREM

First, a number of assumptions are stated for later use. The range of the superscript k for which these assumptions hold is to be specified later.

Assumption 1. $A^k = A^{k,1} + A^{k,2}$ with $A^{k,1}$ symmetric and having eigenvalues $\in (0,B_3.4^k)$, $||A^{k,2}||_{0,k} \le B_4.2^k$.

By $B_1, B_2, \ldots$ we will denote positive constantes that do not depend on the indices k and ℓ. The matrix norm $||A^k||_{0,k}$ is the norm induced by the vector norm $||u^k||_{0,k}$; the use of the same notation for these two norms will not cause confusion.

Assumption 2. A^{k-1} is such that $A^{k-1}u^{k-1} = r^k f^k$ has for all f^k a unique solution which satisfies:

$$||u^k - p^k u^{k-1}||_{0,k} \le B_5 4^{-k} ||f^k||_{0,k},$$

with u^k the solution of $A^k u^k = f^k$.

Before stating a few more assumptions it is convenient to introduce the following definitions:

$$V_1^{k,\gamma} \equiv \{\text{span of all eigenvectors of } A^{k,1} \text{ belonging to eigenvalues } \in (0,\gamma B_3.4^k),$$
$$0 < \gamma < 1,\ \gamma \text{ independent of } k\ ; \tag{3.1}$$

$V_2^{k,\gamma} \equiv$ the orthogonal complement of $V_1^{k,\gamma}$ in U^k;

$$\varepsilon^{k,\mu} \equiv u^{k,\mu} - u^k,\ \mu = 0,\tfrac{1}{2},1,\tfrac{3}{2},\ldots, \tag{3.2}$$

with $u^k \equiv (A^k)^{-1} f^k$ and $u^{k,\mu}$ defined in algorithm 1.

Assumption 3. M^k has the following properties: let $\varepsilon^{k,\mu} = \varepsilon_1^{k,\mu} + \varepsilon_2^{k,\mu}$ with $\varepsilon_i^{k,\mu} \in V_i^{k,\gamma}$, i = 1,2. Then

$$\varepsilon^{k,\mu+\frac{1}{2}} = \sum_{i=1}^{3} \varepsilon_i^{k,\mu+\frac{1}{2}},$$

with $\varepsilon_i^{k,\mu+\frac{1}{2}} \in V_i^{k,\gamma}$, i = 1,2 ;

$$||\varepsilon_1^{k,\mu+\frac{1}{2}}||_{0,k} \le ||\varepsilon_1^{k,\mu}||_{0,k},$$

$$||\varepsilon_2^{k,\mu+\frac{1}{2}}||0,k \le \theta_\gamma ||\varepsilon_2^{k,\mu}||_{0,k},\ \theta_\gamma < 1 \text{ independent of } k,$$

$$||\varepsilon_3^{k,\mu+\frac{1}{2}}||_{0,k} \le (\exp(B_6 2^{-k})-1)||\varepsilon^{k,\mu}||_{0,k} .$$

<u>Assumption 4</u>. $B_7||u^{k-1}||_{0,k-1} \le ||p^k u^{k-1}||_{0,k} \le B_8||u^{k-1}||_{0,k-1}, \forall\, u^{k-1} \in U^{k-1}$.

Define: $\zeta_k \equiv B_5[\{B_3(\gamma+\theta_\gamma) + \exp(B_6.2^{-k})-1\} + B_4.2^{-k}]$,

$$\eta_k \equiv B_7^{-1}B_8\{\exp(B_6.2^{-k}) + \zeta_k\}. \tag{3.3}$$

<u>Assumption 5</u>. There exist integers j and σ such that

$$(\zeta_{j+1} + \eta_{j+1}\zeta_{j+1}^{\sigma-1})^\sigma \le \zeta_{j+1}^{\sigma-1} < 1$$

The essence of the philosophy behind the multiple grid approach is contained in assumptions 2 and 3. Assumption 2 states how well the coarser grid operator A^{k-1} should approximate A^k; assumption 3 states that the relaxation process M^k should annihilate short wavelength components of the error fast, while not amplifying too much long wavelength components, which are to be treated on coarser grids.

Given assumptions 1-5, convergence proofs for algorithms 1 and 2 can be short. Define:

$$\tilde{\delta}_k \equiv \eta_k\delta_{k-1} + \zeta_k. \tag{3.4}$$

<u>Theorem 3.1</u> If assumptions 1-4 are satisfied for k = ℓ then algorithm 1 has the following property:

$$||\varepsilon^{\ell,\mu+1}||_{0,\ell} \le \tilde{\delta}_\ell||\varepsilon^{\ell,\mu}||_{0,\ell},$$

with $\tilde{\delta}_\ell$ defined by (3.4) with k = ℓ.

<u>Proof</u> According to algorithm 1 $f^{\ell-1} = -r^\ell A^\ell \varepsilon^{\ell,\mu+\frac{1}{2}}$. From assumption 2 it follows that

$$||\varepsilon^{\ell,\mu+\frac{1}{2}} + p^\ell u^{\ell-1}||_{0,\ell} \le B_5.4^{-\ell}||A^\ell\varepsilon^{\ell,\mu+\frac{1}{2}}||_{0,\ell}. \tag{3.5}$$

With the use of the identity $\varepsilon^{\ell,\mu+1} = \varepsilon^{\ell,\mu+\frac{1}{2}} + p^\ell u^{\ell-1} + p^\ell(v^{\ell-1}-u^{\ell-1})$ (with $v^{\ell-1}$ an approximation to $u^{\ell-1}$ satisfying (2.2), as required by algorithm 1) it follows from (3.5) and assumption 4 that

$$||\varepsilon^{\ell,\mu+1}||_{0,\ell} \le B_5.4^{-\ell}||A^\ell\varepsilon^{\ell,\mu+\frac{1}{2}}||_{0,\ell} + B_8||v^{\ell-1}-u^{\ell-1}||_{0,\ell-1} .$$

Hence, using (2.2) and assumption 4,

$$||\varepsilon^{\ell,\mu+1}||_{0,\ell} \le B_5.4^{-\ell}||A^\ell\varepsilon^{\ell,\mu+\frac{1}{2}}||_{0,\ell} + \delta_{\ell-1}B_7^{-1}B_8||p^\ell u^{\ell-1}||_{0,\ell} . \tag{3.6}$$

From (3.5) a bound on $||p^\ell u^{\ell-1}||_{0,\ell}$ may be deduced; introduction of this bound in (3.6) results in:

$$||\varepsilon^{\ell,\mu+1}||_{0,\ell} \leq \delta_{\ell-1}B_7^{-1}B_8||\varepsilon^{\ell,\mu+\frac{1}{2}}||_{0,\ell} + B_5.4^{-\ell}(1+\delta_{\ell-1}B_7^{-1}B_8)||A^{\ell}\varepsilon^{\ell,\mu+\frac{1}{2}}||_{0,\ell}. \quad (3.7)$$

Because $\varepsilon_i^{k,\mu+\frac{1}{2}} \perp \varepsilon_2^{k,\mu+\frac{1}{2}}$ one has $||\sum_{i=1}^{2} \varepsilon_i^{k,\mu+\frac{1}{2}}||_{0,k}^2 = \sum_{i=1}^{2} ||\varepsilon_i^{k,\mu+\frac{1}{2}}||^2 \leq ||\varepsilon^{k,\mu}||_{0,k}^2$,

where assumption 3 has been used. Hence

$$||\varepsilon^{\ell,\mu+\frac{1}{2}}||_{0,\ell} \leq ||\varepsilon^{\ell,\mu}||_{0,\ell} + (\exp(B_6.2^{-\ell})-1)||\varepsilon^{\ell,\mu}||_{0,\ell}$$

$$\leq \exp(B_6.2^{-\ell})||\varepsilon^{\ell,\mu}||_{0,\ell}. \quad (3.8)$$

From assumptions 1 and 3 it follows that

$$||A^{\ell}\varepsilon^{\ell,\mu+\frac{1}{2}}||_{0,\ell} \leq \sum_{i=1}^{3} ||A^{\ell,1}\varepsilon_i^{\ell,\mu+\frac{1}{2}}||_{0,\ell} + ||A^{\ell,2}\varepsilon^{\ell,\mu+\frac{1}{2}}||_{0,\ell}$$

$$\leq \{B_3.4^{\ell}(\gamma+\theta_\gamma+\exp(B_6.2^{-\ell})-1) + B_4.2^{\ell}\}||\varepsilon^{\ell,\mu}||_{0,\ell}. \quad (3.9)$$

Substitution of (3.8) and (3.9) in (3.7) completes the proof.

For algorithm 2 we have the following theorem.

Theorem 3.2 If assumptions 1-4 are satisfied for $k = \ell(-1)j+1$ and if assumption 5 is satisfied then algorithm 2 has the following property:

$$||\varepsilon^{\ell,\mu+\sigma}||_{0,\ell} \leq \zeta_{j+1}^{\sigma-1}||\varepsilon^{\ell,\mu}||_{0,\ell}.$$

Proof. From the construction of algorithm 2 it follows, using theorem 1, that

$$||\varepsilon^{k,\mu+\sigma}||_{0,k} \leq \tilde{\delta}_k^{\sigma}||\varepsilon^{k,\mu}||_{0,k}, \qquad \ell > k \geq j+1,$$

with $\tilde{\delta}_k$ defined by (3.4). Hence $\delta_k \leq (\eta_k\delta_{k-1}+\zeta_k)^{\sigma}$. We will show that $\delta_k < \zeta_{j+1}^{\sigma-1}$, $\ell > k \geq j+1$. Since $\delta_j = 0$ we have $\delta_{j+1} \leq \zeta_{j+1}^{\sigma} < \zeta_{j+1}^{\sigma-1}$. Proceeding by induction, if $\delta_k \leq \zeta_{j+1}^{\sigma-1}$ then $\delta_{k+1} < (\eta_{j+1}\zeta_{j+1}^{\sigma-1}+\zeta_{j+1})^{\sigma}$, since $\eta_k < \eta_{j+1}$ and $\zeta_k < \zeta_{j+1}$ for $k > j+1$. Using assumption 5 we obtain $\delta_{k+1} < \zeta_{j+1}^{\sigma-1}$. The proof is concluded by noting that $\tilde{\delta}_{\ell}^{\sigma} < \zeta_{j+1}^{\sigma-1}$.

4. SOME PROPERTIES OF A FINITE DIFFERENCE SCHEME AND OF PROLONGATION AND RESTRICTION OPERATORS.

In this section some results are gathered that are useful for the verification of assumptions 1-4 for the application to be presented in section 5. A specific difference scheme A^{ℓ} is chosen, and matrices A^k, prolongations p^k and restrictions r^k to be used in algorithm 2 are defined.

On Ω^k finite difference operators Δ_i^k and ∇_i^k, $i = 1,2$ are defined as follows:

$$(\Delta_1^k u^k)_{ij} \equiv (u_{i+1,j}^k - u_{ij}^k).2^k, \quad (\nabla_1^k u^k)_{ij} \equiv (u_{ij}^k - u_{i-1,j}^k).2^k, \quad (4.1)$$

and analogous for the x_2-direction. Let $u^k \in U^k$. Then Δ_i^k and ∇_i^k, i = 1,2 are defined for all points of Ω^k. Note that in general $\nabla_i^k u^k$ and $\Delta_i^k u^k \notin U^k$.

The difference scheme that will be considered is defined as follows:

$$A^\ell u^\ell \equiv -\tfrac{1}{2}(\nabla_i^\ell a_{ij} \Delta_j^\ell + \Delta_i^\ell a_{ij} \nabla_j^\ell) u^\ell - \tfrac{1}{2}(\nabla_i^\ell + \Delta_i^\ell)(b_i u^\ell) + c u^\ell = f^\ell. \tag{4.2}$$

In addition to the norm defined by (1.7), three other norms are defined by:

$$||u^k||_{1,k}^2 \equiv \sum_{i=1}^{2} ||\Delta_i^k u^k||_{0,k}^2 \; ; \; ||u^k||_{2,k}^2 \equiv \sum_{i,j=1}^{2} ||\Delta_i^k \nabla_j^k u^k||_{0,k}^2 \; ;$$

$$||u^k||_{-1,k} \equiv \sup_{||v^k||_{1,k} \le 1} |(u^k, v^k)_k| . \tag{4.3}$$

One easily verifies that

$$||\nabla_i^k u^k||_{0,k} = ||\Delta_i^k u^k||_{0,k} . \tag{4.4}$$

The derivation of the following partial summation formulae is trivial:

$$(\nabla_i^k u^k, v^k)_k = -(u^k, \Delta_i^k v^k)_k , (\nabla_i^k u^k, \nabla_i^k v^k) = -(\nabla_i^k \Delta_i^k u^k, v^k) , \tag{4.5}$$

A number of lemmata that will be used in the sequel are presented.

Lemma 4.1. $||u^k||_{0,k} \le ||\Delta_i^k u^k||_{0,k}$, i = 1,2.

Proof. This is a discrete version of Poincaré's inequality, and the method of proof is well-known.

Lemma 4.2. $||u^k||_{-1,k} \le ||u^k||_{0,k} \le ||u^k||_{1,k} \le ||u^k||_{2,k}$.

Lemma 4.3. $|(u^k,v^k)_k| \le \frac{1}{2\varepsilon} (u^k,u^k)_k + \frac{\varepsilon}{2} (v^k,v^k)_k$, $\forall \, \varepsilon > 0$.

Lemma 4.4. $||u^k||_{1,k} \le \frac{1}{2\varepsilon} ||u^k||_{0,k} + \frac{\varepsilon}{2} ||u^k||_{2,k}$, $\forall \, \varepsilon > 0$.

The proofs of the lemmata 4.2, 4.3 are elementary. Lemma 4.4 follows from (4.5) and lemma 4.3.

We proceed to establish a number of results that will be used to prove the existence and uniqueness of the solution of (4.2) and a property of its $||.||_{2,\ell}$ norm. The following considerations lean strongly on [4]. The reason why we cannot just quote [4] is that in [4] a slightly different difference scheme is studied. The difference lies in the discretisation of the mixed derivative $u_{,12}$. In [4] this is done as follows:

$$u_{,12} \cong \frac{1}{4} (\nabla_1^\ell + \Delta_1^\ell)(\nabla_2^\ell + \Delta_2^\ell) u^\ell , \tag{4.6}$$

whereas in the present case, if $a_{12} \equiv 1$ we would have:

$$u_{,12} = \tfrac{1}{2}(\nabla_1^\ell \Delta_2^\ell + \Delta_1^\ell \nabla_2^\ell)u^\ell. \tag{4.7}$$

The use of (4.7) rather than (4.6) is essential for the derivation of certain properties of the bilinear form $B(.,.)$ to be introduced later. A difference similar to that between (4.6) and (4.7) occurs in the definition of $||.||_{2,k}$ in [4] and here. Apart from this difference, lemma 4.4 occurs also in [3].

Define:

$$A^{\ell,3} \equiv c,\ A^{\ell,4} \equiv A^\ell - A^{\ell,3} - A^{\ell,5},\quad A^{\ell,5} \equiv -\tfrac{1}{2}a_{ij}(\nabla_i^\ell \Delta_j^\ell + \Delta_i^\ell \nabla_j^\ell). \tag{4.8}$$

It is not difficult to show that

$$||A^{\ell,4}u^\ell||_{0,\ell} \leq (\tilde{\alpha}^2+\beta^2)^{\frac{1}{2}}||u^\ell||_{1,\ell} \tag{4.9}$$

with

$$\tilde{\alpha}^2 \equiv \sum_{j=1}^{2} \sup_\Omega |\nabla_i a_{ij}|^2,\ \beta^2 \equiv \sup_\Omega b_i b_i. \tag{4.10}$$

Lemma 4.5. $||u^\ell||_{2,\ell} \leq B_9||A^{\ell,5}u^\ell||_{0,\ell}$ with $B_9 \equiv B_2\sqrt{2}/B_1^2$.

A similar result is deduced in [14] (p.296, eq. (18)) with a slightly different definition of $||.||_{2,\ell}$ and $A^{\ell,5}$. The proof presented in [14] also holds for the present case with some minor modifications.

Define

$$\gamma \equiv \sup_\Omega |c|. \tag{4.11}$$

Lemma 4.6. $||u^\ell||_{2,\ell} \leq 2B_9\{||A^\ell u^\ell||_{0,\ell} + B_{10}||u^\ell||_{0,\ell}\}$ with $B_{10} \equiv (\alpha^2+\beta^2)B_9/2+\gamma$.

This lemma is easily derived using lemma 4.5, (4.8), (4.9) and lemma 4.4 with $\varepsilon = (\tilde{\alpha}^2+\beta^2)^{-\frac{1}{2}}B_9^{-1}$.

Theorem 4.1. If (1.1) has a unique (weak) solution and ℓ is large enough then there exists a constant $B_{11} > 0$ independent of ℓ such that

$$||A^\ell u^\ell||_{0,\ell} \geq B_{11}||u^\ell||_{0,\ell}.$$

The proof uses lemma 4.6 and is given in [4], pp. 106, 107 with a slightly different definition of A^ℓ as discussed after lemma 4.4, but is easily seen to hold also in the present case; lemma 4.6 plays an important role in the proof.

Theorem 4.2. If theorem 4.1 holds then $||u^\ell||_{2,\ell} \leq B_{12}||A^\ell u^\ell||_{0,\ell}$ with $B_{12} \equiv 2B_9(1+B_{10}/B_{11})$.

This theorem follows directly from theorem 4.1 and lemma 4.6.

Define the following bilinear form:

$$B^{\ell}(u^{\ell},v^{\ell}) \equiv (A^{\ell}u^{\ell},v^{\ell})_{\ell}\ , \quad u^{\ell},v^{\ell} \in U^{\ell}. \tag{4.12}$$

The form B^{ℓ} will play an important role in the subsequent analysis. We proceed to establish a number of useful properties.
Application of (4.5) yields:

$$B^{\ell}(u^{\ell},v^{\ell}) = \tfrac{1}{2}(a_{ij}\Delta_j^{\ell}u^{\ell},\Delta_i^{\ell}v^{\ell})_{\ell} + \tfrac{1}{2}(a_{ij}\nabla_j^{\ell}u^{\ell},\nabla_i^{\ell}v^{\ell})_{\ell}$$

$$+ \tfrac{1}{2}(b_i u^{\ell},\Delta_i^{\ell}v^{\ell}+\nabla_i^{\ell}v^{\ell})_{\ell} + (cu^{\ell},v^{\ell})_{\ell}. \tag{4.13}$$

<u>Lemma 4.7</u>. For every $u^{\ell} \in U^{\ell}$ there exists an element $z^{\ell} \in U^{\ell}$ such that

(i) $B^{\ell}(u^{\ell},u^{\ell}+z^{\ell}) \geq \tfrac{1}{2}B_1||u^{\ell}||^2_{1,\ell}$;

(ii) $||z^{\ell}||_{2,\ell} \leq B_{13}||u^{\ell}||_{0,\ell}$,

with $B_{13} \equiv B_{12} \sup_{\Omega} |\beta^2/2B_1 - c|$.

<u>Proof</u> Define $z^{\ell} \in U^{\ell}$ by:

$$A^{\ell *} z^{\ell} = (\mu - c)u^{\ell}, \tag{4.14}$$

with $A^{\ell *}$ the adjoint of A^{ℓ} and $\mu > 0$ a constant to be determined later. From theorem 4.1 existence and uniqueness of z^{ℓ} follow, and from theorem 4.2 it follows that

$$||z^{\ell}||_{2,\ell} \leq B_{12} \sup_{\Omega} |\mu - c|\ ||u^{\ell}||_{0,\ell}. \tag{4.15}$$

The remainder of the proof is quite similar to the reasoning in [3], pp. 128-130. In the proof the choice $\mu = \beta^2/2B_1$ is made.

<u>Theorem 4.3</u>.
(i) $|B^{\ell}(u^{\ell},v^{\ell})| \leq B_{14}||u^{\ell}||_{1,\ell}||v^{\ell}||_{1,\ell}$

with $B_{14} \equiv \alpha+\beta+\gamma$, $\alpha \equiv (\sup_{\Omega} a_{ij}a_{ij})^{\frac{1}{2}}$, β and γ defined by (4.10) and (4.11);

(ii) $\sup_{||v^{\ell}||_{1,\ell} \leq 1} |B^{\ell}(u^{\ell},v^{\ell})| \geq B_{15}||u^{\ell}||_{1,\ell}$, $B_{15} \equiv \tfrac{1}{2}B_1(1+B_{13})^{-1}$;

(iii) $\sup_{||u^{\ell}||_{1,\ell} \leq 1} |B^{\ell}(u^{\ell},v^{\ell})| \geq B_{15}||v^{\ell}||_{1,\ell}$.

<u>Proof</u>. Cauchy-Schwarz and (4.17) yield:

$$|B^\ell(u^\ell,v^\ell)| \le \sup_\Omega |a_{ij}|\; ||\Delta_i^\ell u^\ell||_{0,\ell}||\Delta_j^\ell v^\ell||_{0,\ell} + \sup_\Omega |b_i|\; ||u^\ell||_{0,\ell}||\Delta_i^\ell v^\ell||_{0,\ell}$$

$$+ \gamma||u^\ell||_{0,\ell}||v^\ell||_{0,\ell} .$$

Lemma 4.1 and repeated use of Hölder's inequality result in (i). We continue with (ii). Let $v^\ell = (u^\ell+z^\ell)||u^\ell+z^\ell||_{1,\ell}^{-1}$, with z^ℓ as in lemma 4.7. Then obviously $||v^\ell||_{1,\ell} = 1$, and $B^\ell(u^\ell,v^\ell) \ge \frac{1}{2}B_1||u^\ell||_{1,\ell}^2||u^\ell+z^\ell||_{1,\ell}^{-1}$.
Furthermore, $||u^\ell+z^\ell||_{1,\ell} \le ||u^\ell||_{1,\ell} + ||z^\ell||_{2,\ell} \le (1+B_{13})||u^\ell||_{1,\ell}$ because of (ii) of lemma 4.7, and (ii) is established. By reversing the roles of u^ℓ and v^ℓ one may similarly prove (iii).

Lemma 4.7 and theorem 4.3 are related to results in [3], pp. 125-130 concerning the differential equation (1.1).

We now proceed to construct bilinear forms $B^k(u^k,v^k)$, $k < \ell$ that can be used to approximate B^ℓ and each other in a sense that will become clear in the sequel. These bilinear forms are related to the operators A^k needed in algorithms 1 and 2.

It has already been assumed in section 2 that there exist prolongation operators $p^k : U^{k-1} \to U^k$ and restriction operators $r^k : U^k \to U^{k-1}$. Furthermore, define prolongations $p^{mk} : U^k \to U^m$ and restrictions $r^{km} : U^m \to U^k$, $k < m$ as follows:

$$p^{mk} \equiv p^m p^{m-1} \dots p^{k+1} , \quad r^{km} \equiv r^{k+1} r^{k+2} \dots r^m \tag{4.16}$$

The operators p^k and r^k have to satisfy assumption 4 and must have the following properties:

$$(p^k u^{k-1},v^k)_k = (u^{k-1},r^k v^k)_{k-1} , \quad \forall\, v^k \in U^k , \tag{4.17a}$$

$$||u^k-p^k r^k u^k||_{1,k} \le B_{16}.2^{-k}||u^k||_{2,k} , \tag{4.17b}$$

$$||u^k-r^{k+1}p^{k+1}u^k||_{-1,k} \le B_{17}.4^{-k}||u^k||_{1,k} , \tag{4.17c}$$

$$||r^k u^k||_{0,k-1} \le B_{18}||u^k||_{0,k} , \tag{4.17d}$$

$$||u^k||_{2,k} \le B_{19}||p^{k+1}u^k||_{2,k+1} , \tag{4.17e}$$

$$||r^k u^k||_{2,k-1} \le B_{20}||u^k||_{2,k} , \tag{4.17f}$$

$$B_{21}||u^k||_{1,k} \le ||p^{mk}u^k||_{1,m} \le B_{22}||u^k||_{1,k} , \quad m > k . \tag{4.17g}$$

Useful consequences of (4.17) are:

$$(p^{mk}u^k,v^m)_m = (u^k,r^{km}v^m)_k$$

which follows from (4.17). Hence

$$||r^{km}u^m||_{-1,k} = \sup_{||v^k||_{1,k}\le 1} |(r^{km}u^m,v^k)_k| = \sup_{||v^k||_{1,k}\le 1} |(u^m,p^{mk}v^k)_m|$$

$$\le \sup_{||p^{mk}v^k||_{1,m}\le B_{22}} |(u^m,p^{mk}v^k)_m| \le B_{22} \sup_{||v^m||_{1,m}\le 1} |(u^m,v^m)_m| = B_{22}||u^m||_{-1,m} \tag{4.18}$$

where (4.17g) has been used. Furthermore we have:

$$||u^m-p^{mk}r^{km}u^m||_{1,m} \le B_{1km}||u^m||_{2,m}\ , \ k < m\ , \tag{4.19a}$$

with

$$B_{1km} \equiv 2^{-m}B_{16}\{1+2B_{20}B_{22}((2B_{20})^{m-k-1}-1)(2B_{20}-1)^{-1}\},$$

$$||u^k-r^{km}p^{mk}u^k||_{-1,m} \le B_{2km}||u^k||_{1,k}\ , \ k < m\ , \tag{4.19b}$$

with

$$B_{2km} \equiv 4^{-k}B_{17}\{1+\frac{1}{3}B_{22}^2(1-4^{k-m+1})\}.$$

Equation (4.19a) may be derived by noting that

$$u^m-p^{mk}r^{km}u^m = (1-p^mr^m)u^m + p^m(1-p^{m-1}r^{m-1})r^mu^m + \dots$$
$$+ p^{m,k+1}(1-p^{k+1}r^{k+1})r^{k+1,m}u^m\ ,$$

and by using (4.17b, f, g). The derivation of (4.19b) is similar.

For example, p^k and r^k defined below satisfy assumption 4 and (4.17):

$$(p^ku^{k-1})_{ij} = \begin{cases} u^{k-1}_{i/2,j/2}\ , \ i \text{ and } j \text{ even}\ ; \\ \frac{1}{2}(u^{k-1}_{(i+1)/2,j/2} + u^{k-1}_{(i-1)/2,j/2}\ , \ i \text{ odd, } j \text{ even}\ , \\ \frac{1}{2}(u^{k-1}_{i/2,(j+1)/2} + u^{k-1}_{i/2,(j-1)/2}\ , \ i \text{ even, } j \text{ odd}\ ; \\ \frac{1}{4}(u^{k-1}_{(i+1)/2,(j+1)/2} + u^{k-1}_{(i+1)/2,(j-1)/2} + u^{k-1}_{(i-1)/2,(j+1)/2} \\ \quad + u^{k-1}_{(i-1)/2,(j-1)/2}),\ i \text{ odd, } j \text{ odd.} \end{cases} \tag{4.20a}$$

This p^k corresponds to linear interpolation. Once p^k has been chosen, r^k is fixed by (4.17a). One obtains:

$$(r^ku^k)_{ij} = \frac{1}{4}u^k_{2i,2j} + \frac{1}{8}(u^k_{2i+1,2j} + u^k_{2i-1,2j} + u^k_{2i,2j+1} + u^k_{2i,2j-1})$$
$$+ \frac{1}{16}(u^k_{2i+1,2j+1} + u^k_{2i+1,2j-1} + u^k_{2i-1,2j+1} + u^k_{2i-1,2j-1}) \tag{4.20b}$$

It is not difficult to verify, that for p^k and r^k as defined by (4.20) assumption

4 and conditions (4.17) are satisfied with

$$B_7=1/2,\ B_8=1,\ B_{16}=13/16,\ B_{17}=3/16,\ B_{18}=B_{19}=B_{20}=1,\ B_{21}=3^{-1/2},\ B_{22}=1. \tag{4.21}$$

The constant that is most difficult to determine is B_{21}. It is found by noting that

$$||\nabla_1^m p^{mk} u^k||^2_{0,m} = 2^{k-m}||\nabla_1^k u^k||^2_{0,k} + 2^{k-m-2} \sum_{i,j=0}^{2^k} \sum_{p=1}^{n-1} \left(\frac{2n-p}{n} w^k_{ij} + \frac{p}{n} w^k_{i,j+1}\right)^2$$

with $w^k = \nabla_1^k u^k$, $n = 2^{m-k-1}$, and by using the inequality

$$\Sigma(a_i+b_i)^2 \geq \Sigma(a_i^2+b_i^2) - 2(\Sigma a_i^2)^{\frac{1}{2}}(\Sigma b_i^2)^{\frac{1}{2}} .$$

We now define A^k as follows:

$$A^{k-1} = r^k A^k p^k \ , \ k = \ell,\ell-1,\ell-2,\ldots \tag{4.22}$$

The equation $A^k u^k = f^k$ is equivalent to

$$B^k(u^k,v^k) = (f^k,v^k)_k,\ \forall\ v^k \in U^k \tag{4.23}$$

with B^k defined by $B^k(u^k,v^k) = (A^k u^k,v^k)$, from which it follows that

$$B^k(u^k,v^k) = B^\ell(p^{\ell k}u^k,p^{\ell k}v^k). \tag{4.24}$$

For these results (4.17a) is essential. In order to verify whether A^k satisfies assumption 2 of section 3 the properties of B^k will be studied. For future use the following constants are defined:

$$\left.\begin{aligned}
&B_{1k} \equiv (\tfrac{1}{2}B_1-B_{13}B_{14}B_{1k\ell})B_{21}^2,\ B_{2k} \equiv B_{13}(B_8B_{20})^{\ell-k} ,\\
&B_{3k} \equiv B_{1k}(1+B_{2k})^{-1} ,\\
&B_{4k} \equiv (B_8B_{20})^{\ell-k}B_{12} + 2^{k+1}\{B_{3k}^{-1}B_{2k\ell}2^{k+1}\\
&\qquad + B_8^{\ell-k}B_{12}B_{21}^{-1}((1+B_{14}B_{22}^2/B_{3k})\delta_{k\ell}+B_{1k\ell})\} ,\\
&B_{3mn} \equiv B_{14}B_{22}^2B_{4n}(1+B_{14}B_{22}^2/B_{3m}),
\end{aligned}\right\} \tag{4.25}$$

with $B_{1k\ell}$ and $B_{2k\ell}$ defined after equations (4.19a and b), and $\delta_{k\ell}$ a number such that (4.26) below is satisfied.

The following theorems will be proved:

<u>Theorem 4.4.</u> For k = 1(1) -1 the following assertions hold:

(i) $|B^k(u^k,v^k)| \leq B_{14}B_{22}^2||u^k||_{1,k}||v^k||_{1,k}$,

(ii) $$\sup_{||v^k||_{1,k}\le 1} |B^k(u^k,v^k)| \ge B_{3k}||u^k||_{1,k} ,$$

(iii) $$\sup_{||u^k||_{1,k}\le 1} |B^k(u^k,v^k)| \ge B_{3k}||v^k||_{1,k} .$$

This generalization of theorem 4.3 enables us to use the following theorem:

Theorem 4.5. Let $B^k(u^k,v^k)$ satisfy theorem 4.4 and assume $B_{3k} > 0$ for $k \ge j$. Then $u^m \in U^m$ and $u^n \in U^n$, $j \le m < n \le \ell$ defined by the following equations:

$$B^m(u^m,v^m) = (r^{mn}f^n,v^m)_m, \ \forall\, v^m \in U^m ,$$

$$B^n(u^n,v^n) = (f^n,v^n)_n, \ \forall\, v^n \in U^n ,$$

exist and are unique. Furthermore, suppose that for an element $w^m \in U^m$ and a number δ_{mn} we have:

$$||p^{nm}w^m-u^n||_{1,n} \le \delta_{mn}||u^n||_{2,n} \tag{4.26}$$

Then

$$||p^{nm}u^m-u^n||_{1,n} \le (1+B_{14}B_{22}^2/B_{3m})\delta_{mn}||u^n||_{2,n}. \tag{4.27}$$

Corollary. $||u^m||_{2,m} \le B_{19}^{n-m}\{1+(1+B_{14}B_{22}^2/B_{3m})2^{n+1}\delta_{mn}\}||u^n||_{2,n}$

Theorem 4.5 gives information about the discrepancy between B^m and B^n and hence between A^m and A^n. The quantity $||u^k||_{2,k}$ occurs in the error estimate; the corollary enables us to obtain some information on $||u^k||_{2,k}$ on coarser grids. Theorem 4.5 is a direct application of theorem 2.2 of [2], p.324. It turns out that theorem 4.5 is not quite what we need for the verification of assumption 2 of section 2, but that we need the following modified version of theorem 4.5:

Theorem 4.6. Let the assumptions of theorem 4.5 be fulfilled, and let there exist for $\forall\, u^n \in U_n$ a w^n such that (4.26) holds. Then for $j \le m < n \le \ell$ we have:

$$||p^{nm}u^m-u^n||_{0,n} \le B_{3mn}\delta_{mn}^2||u^n||_{2,n} .$$

We proceed to prove these theorems. The following lemmata are helpful.

Lemma 4.8. For $k < \ell$:

$$|B^k(u^k,v^k)| \le B_{14}B_{22}^2||u^k||_{1,k}||v^k||_{1,k} .$$

This lemma follows directly from (4.24), theorem 4.3 (i) and (4.17g). The following lemma generalizes lemma 4.7:

Lemma 4.9. For $k \le \ell-1$ and $\forall\, u^k \in U^k$ there exists an element $z^k \in U^k$ such that:

(i) $B^k(u^k,u^k+z^k) \geq B_{1k}||u^k||^2_{1,k}$

(ii) $||z^k||_{2,k} \leq B_{2k}||u^k||_{0,k}$

Proof. The following inequality holds for $\forall\, w^k \in U^k$ and $\forall\, v^\ell \in U^\ell$:

$$B^k(u^k,w^k) \geq B^\ell(p^{\ell k}u^k,v^\ell) - |B^\ell(p^{\ell k}u^k,p^{\ell k}w^k-v^\ell)|,$$

where (4.24) has been used. From lemma 4.7 it follows that there exists an element $z^\ell \in U^\ell$ such that

$$B^\ell(p^{\ell k}u^k,p^{\ell k}u^k+z^\ell) \geq \tfrac{1}{2}B_1||p^{\ell k}u^k||^2_{1,\ell}$$

and

$$||z^\ell||_{2,\ell} \leq B_{13}||p^{\ell k}u^k||_{0,\ell}.$$

Choose $v^\ell = p^{\ell k}u^k + z^\ell$. Then, using theorem 4.3 (i),

$$B^k(u^k,w^k) \geq \tfrac{1}{2}B_1||p^{\ell k}u^k||^2_{1,\ell} - B_{14}||p^{\ell k}u^k||_{1,\ell}||p^{\ell k}w^k-v^\ell||_{1,\ell}.$$

Choose $w^k = u^k + z^k$, $z^k = r^{k\ell}z^\ell$. Then, using (4.17f) and lemma 4.7 (ii), z^k satisfies (ii) above. Furthermore, $p^{\ell k}w^k - v^\ell = p^{\ell k}r^k z^\ell - z^\ell$.
Hence, using (4.19a),

$$||p^{\ell k}w^k - v^\ell||_{1,\ell} \leq B_{1k\ell}||z^\ell||_{2,\ell}.$$

Using lemma 4.7 (ii) and lemma 4.2 one obtains:

$$B^k(u^k,w^k) \geq (\tfrac{1}{2}B_1-B_{13}B_{14}B_{1k\ell})||p^{\ell k}u^k||^2_{1,\ell},$$

and application of (4.17g) completes the proof.

Theorem 4.4 can be proved as follows. Assertion (i) is identical to lemma 4.8. Choosing $v^k = (u^k+z^k)||u^k+z^k||^{-1}_{1,k}$ with z^k as in lemma 4.9 it follows from lemma 4.9 that

$$B^k(u^k,v^k) \geq B_{1k}||u^k||^2_{1,k}||u^k+z^k||^{-1}_{1,k}.$$

Since $||u^k+z^k||_{1,k} \leq ||u^k||_{1,k}(1+B_{2k})$ (lemmata 4.9 (ii) and 4.2) (ii) follows. By reversing the roles of u^k and v^k one obtains (iii), and the proof of theorem 4.4 is complete. Note that B_{3k} is not necessarily positive, but that it will be positive for k sufficiently large but independent of ℓ.

The proof of theorem 4.5 will not be given since, as has already been remarked, it is a direct application of theorem 2.2 of [2], p.324.
The corollary of theorem 4.5 is proved as follows. Since $||u^k||_{2,k} \leq 2^{k+1}||u^k||_{1,k}$, $\forall\, u^k \in U^k$ it follows from theorem 4.5 that

$$||p^{nm}u^m - u^n||_{2,n} \leq (1+B_{14}B_{22}^2/B_{3m})2^{n+1}\delta_{mn}||u^n||_{2,n} ,$$

hence

$$||p^{nm}u^m||_{2,n} \leq \{1 + (1+B_{14}B_{22}^2/B_{3m})2^{n+1}\delta_{mn}\}||u^n||_{2,n} .$$

With repeated use of (4.17e) the corollary follows.

Before proceeding with theorem 4.6 the following theorem will be established:

Theorem 4.7. Assume that for a given value of $k < \ell$ theorem 4.4 holds with $B_{3k} > 0$. Then the solution u^k of the equation

$$B^k(u^k,v^k) = (f^k,v^k)_k, \forall v^k \in U^k$$

satisfies:

(i) $||u^k||_{1,k} \leq B_{3k}^{-1}||f^k||_{-1,k}$,

(ii) $||u^k||_{2,k} \leq B_{4k}||f^k||_{0,k}$.

Proof. Existence and uniqueness of u^k follow from theorem 4.5.
From theorem 4.4 (ii) one easily obtains (i).
Continuing with (ii), define $\hat{u}^k$ and $\hat{u}^\ell$ as follows:

$$B^k(\hat{u}^k,v^k) = (r^{k\ell}p^{\ell k}f^k,v^k)_k , \forall v^k \in U^k ,$$

$$B^\ell(\hat{u}^\ell,v^\ell) = (p^{\ell k}f^k,v^\ell)_\ell , \forall v^\ell \in U^\ell .$$

Using equations (4.17g), (4.19a), theorem 4.5 and 4.2 and assumption 4 of section 3 one may derive:

$$||\hat{u}^k-r^{k\ell}\hat{u}^\ell||_{1,k} \leq B_{21}^{-1}||p^{\ell k}\hat{u}^k-p^{\ell k}r^{k\ell}\hat{u}^\ell||_{1,\ell}$$

$$\leq B_{21}^{-1}||p^{\ell k}\hat{u}^k-\hat{u}^\ell||_{1,\ell} + B_{21}^{-1}||\hat{u}^\ell-p^{\ell k}r^{k\ell}\hat{u}^\ell||_{1,\ell}$$

$$\leq B_{21}^{-1}\{(1+B_{14}B_{22}^2/B_{3k})\delta_k + B_{1k\ell}\}||\hat{u}^\ell||_{2,\ell}$$

$$\leq B_8^{\ell-k}B_{12}B_{21}^{-1}\{(1+B_{14}B_{22}^2/B_{3k})\delta_k + B_{1k\ell}\}||f^k||_{0,k} . \qquad (4.28)$$

Furthermore,

$$B^k(u^k-\hat{u}^k,v^k) = (f^k-r^{k\ell}p^{\ell k}f^k,v^k)_k , \forall v^k \in U^k .$$

From (i) above and (4.19b) it follows that

$$||u^k-\hat{u}^k||_{1,k} \leq B_{3k}^{-1}B_{2k\ell}||f^k||_{1,k} \leq B_{3k}^{-1}B_{2k\ell}2^{k+1}||f^k||_{0,k} \qquad (4.29)$$

Combination of (4.28) and (4.29) and the use of the inequality

$||u^k||_{2,k} \leq 2^{k+1}||u^k||_{1,k}$ results in:

$$||u^k - r^{k\ell}\hat{u}^\ell||_{2,k} \leq 2^{k+1}\{B_{3k}^{-1}B_{2k\ell}2^{k+1} + B_8^{\ell-k}B_{12}B_{21}^{-1}((1+B_{14}B_{22}^2/B_{3k})\delta_{k\ell} + B_{1k\ell})\}||f^k||_{0,k} \quad (4.30)$$

From (4.17f), theorem 4.2 and assumption 4 of section 3 it follows that

$$||r^{k\ell}\hat{u}^\ell||_{2,k} \leq (B_8B_{20})^{\ell-k}B_{12}||f^k||_{0,k} \; . \quad (4.31)$$

Combination of (4.30) and (4.31) completes the proof of (ii).

Corollary. Assume that for a given value of $k < \ell$ theorem 4.4 holds with $B_{3k} > 0$. Then the solution u^k of the equation

$$B^k(v^k,u^k) = (v^k,f^k)_k \; , \; \forall \; v^k \in U^k$$

satisfies theorem 4.7.

This corollary may be proved by reversing the roles of u^k and v^k in the proof of theorem 4.7.

The following proof of theorem 4.6 is analogous to what is known as "Nitsche's tric" in the finite element literature.

Define $w^n \in U^n$ by $B^n(v^n,w^n) = (v^n,u^n-p^{nm}u^m)_n$, $\forall \; v^n \in U^n$.

Existence and uniqueness of w^n follow from a straightforward application of theorem 2.2 of [1], p.324. One has

$$||u^n-p^{nm}u^m||_{0,n}^2 = B^n(u^n-p^{nm}u^m,w^n).$$

Since $B^n(u^n-p^{nm}u^m,p^{nm}v^m) = 0$, $\forall \; v^m \in U^m$ one may write, using theorem 4.4 (i),

$$||u^n-p^{nm}u^m||_{0,n}^2 = B^n(u^n-p^{nm}u^m,w^n-p^{nm}v^m)$$

$$\leq B_{14}B_{22}^2||u^n-p^{nm}u^m||_{1,n}||w^n-p^{nm}v^m||_{1,n} \; , \; \forall \; v^m \in U^m. \quad (4.32)$$

Choose v^m such that $||w^n-p^{nm}v^m||_{1,n} \leq \delta_{mn}||w^n||_{2,n}$. Using the corollary of theorem 4.7, $||w^n||_{2,n} \leq B_{4n}||u^n-p^{nm}u^m||_{0,n}$, so that

$$||w^n-p^{nm}v^m||_{1,n} \leq \delta_{mn}B_{4n}||u^n-p^{nm}u^m||_{0,n}.$$

Substitution in (4.32) gives

$$||u^n-p^{nm}u^m||_{0,n} \leq B_{14}B_{22}^2B_{4n}\delta_{mn}||u^n-p^{nm}u^m||_{1,n}.$$

Application of theorem 4.5 completes the proof.

5. APPLICATION

We now choose A^k in algorithm 2 as follows:

$$A^{k-1} = r^k A^k p^k \ , \ k = \ell,\ell-1,\ldots, \tag{5.1}$$

with r^k and p^k defined by (4.20). Hence $(A^k u^k,v^k)_k = B^k(u^k,v^k)$, with B^k the bilinear form that has been studied in the preceding section. Assumption 4 is satisfied with $B_9 = \frac{1}{2}$, $B_8 = 1$. We proceed with the verification of assumption 1. Let

$$A^\ell = A^{\ell,1} + A^{\ell,2} \ ,$$

with

$$A^{\ell,1} \equiv -\tfrac{1}{2}\nabla_i^\ell a_{ij}\Delta_j^\ell - \tfrac{1}{2}\Delta_i^\ell a_{ij}\nabla_j^\ell \ , \ A^{\ell,2} \equiv -\tfrac{1}{2}(\nabla_i^\ell+\Delta_i^\ell)b_i + c \ ,$$

and let

$$B^\ell = B^{\ell,1} + B^{\ell,2} \ , \text{ with } B^{\ell,i}(u^\ell,v^\ell) \equiv (A^{\ell,i}u^\ell,v^\ell)_\ell \ , \ i = 1,2.$$

Obviously,

$$B^{\ell,1}(u^\ell,u^\ell) > 0 \ \text{ if } \ u^\ell \neq 0. \tag{5.2}$$

Using the inequality of Cauchy-Schwarz one obtains:

$$|B^{\ell,1}(u^\ell,v^\ell)| \leq \alpha||u^\ell||_{1,\ell}||v^\ell||_{1,\ell} \tag{5.3}$$

(α defined in theorem 4.3), hence

$$|B^{\ell,1}(u^\ell,v^\ell)| \leq B_3.4^\ell||u^\ell||_{0,\ell}||v^\ell||_{0,\ell}$$

with $B_3 \equiv 4\alpha$, so that

$$||A^{\ell,1}||_{0,\ell} \leq B_3.4^\ell. \tag{5.4}$$

Furthermore, $B^{\ell,1}(u^\ell,v^\ell) = B^{\ell,1}(v^\ell,u^\ell)$, hence $A^{\ell,1}$ is self-adjoint, so that, using (5.2) and (5.4),

$$\lambda(A^{\ell,1}) \in (0,B_3.4^\ell], \tag{5.5}$$

if $\lambda(A^{\ell,1})$ is an eigenvalue of $A^{\ell,1}$. Next we observe that

$$|B^{\ell,2}(u^\ell,v^\ell)| \leq \beta||u^\ell||_{0,\ell}||v^\ell||_{1,\ell} + \gamma||u^\ell||_{0,\ell}||v^\ell||_{0,\ell} \tag{5.6}$$

(β and γ defined by (4.10) and (4.11)), so that

$$||A^{\ell,2}||_{0,\ell} \leq B_4.2^\ell \ , \tag{5.7}$$

with $B_4 \equiv 2\beta+\gamma$. From (5.5) and (5.7) it follows that assumption 1 is satisfied for $k = \ell$. The case $k < \ell$ is treated as follows. We have

$$B^{k,1}(u^k,u^k) = B^{\ell,1}(p^{\ell k}u^k,p^{\ell k}u^k) > 0 \ \text{ if } \ u^k \neq 0 \ , \tag{5.8}$$

because of (5.2). From (5.3), (4.17g) and (4.21) it follows that

$$|B^{k,1}(u^k,v^k)| \le \alpha||u^k||_{1,k}||v^k||_{1,k} . \tag{5.9}$$

Furthermore,

$$|B^{k,2}(u^k,v^k)| = |B^{\ell,2}(p^{\ell k}u^k,p^{\ell k}v^k)| \le$$

$$\le \beta||u^k||_{0,k}||v^k||_{1,k} + \gamma||u^k||_{0,k}||v^k||_{0,k} , \tag{5.10}$$

where (5.6), (4.17g), assumption 4 and (4.21) have been used. We conclude that (5.2), (5.3) and (5.6) hold with ℓ replaced by k, and by proceeding as in the case $k = \ell$ one easily verifies that assumption 1 is satisfied.

We proceed with assumption 2. From (4.21) it follows that

$$\left.\begin{aligned} &B_{1km} = 13.2^{-m-4}(2^{m-k}-1) < 13.2^{-4-k} , \\ &B_{2km} = 3.4^{-k-2}(1+3^{-1}(1-4^{k-m+1})) < 4^{-k-1} , \\ &B_{1k} \ge \tfrac{1}{3}(\tfrac{1}{2}B_1-B_{13}B_{14}.13.2^{-4-j}),\ k \ge j,\ B_{2k} = B_{13} , \\ &B_{3k} = B_{1k}(1+B_{13})^{-1}. \end{aligned}\right\} \tag{5.11}$$

Choosing $w^m = r^{mn}u^n$ in (4.34) it follows from (4.19a) and (4.21) that

$$\delta_{mn} = 13.2^{-4-m} . \tag{5.12}$$

Furthermore, if we define

$$B_{23} \equiv B_{12} + B_{3j}^{-1} + \frac{13}{8}3^{1/2}B_{12}(2+B_{14}/B_{3j}) , \tag{5.13}$$

then

$$B_{4k} \le B_{23},\ k \ge j,\ B_{3mn} \le B_{14}B_{23}(1+B_{14}/B_{3j}) ,\ m,n \ge j. \tag{5.14}$$

From now on, let j be the lowest value of k for which $B_{1k} > 0$. From theorem 4.5 it follows that $A^{k-1}u^{k-1} = r^kf^k$ has a unique solution for $k \ge j+1$, which is the first part of assumption 2. From theorems 4.6 and 4.7, and equations (5.12) and (5.14) it follows that the second part of assumption 2 is satisfied with

$$B_5 = B_{14}B_{23}^2(1+B_{14}/B_{3j})(13/8)^2 . \tag{5.15}$$

The iteration process M^k that occurs in algorithms 1 and 2 is chosen as follows:

$$\begin{aligned} &u^{k,\mu,0}:=u^{k,\mu};u^{k,\mu,\nu+1}:=u^{k,\mu,\nu}-\alpha(A^ku^{k,\mu,\nu}-f^k),\nu=0(1)m-1; \\ &u^{k,\mu+\frac{1}{2}}:=u^{k,\mu,m}; \end{aligned} \tag{5.16}$$

with α a parameter to be specified shortly. It will be shown that assumption 3 is satisfied. From (5.16) it follows that

$$\varepsilon^{k,\mu+\frac{1}{2}} = (I^k-\alpha A^k)^m \varepsilon^{k,\mu},$$

with I^k the identity operator. Define:

$$T^{k,1} \equiv I^k-\alpha A^{k,1} \ , \ T^{k,2} \equiv -\alpha A^{k,2} \ .$$

Let $\varepsilon^{k,\mu} = \varepsilon_1^{k,\mu} + \varepsilon_2^{k,\mu}$ with $\varepsilon_i^{k,\mu} \in V_i^{k,\gamma}$, $i = 1,2$, with $V_i^{k,\gamma}$ defined in equation (3.1), (3.2). Then one may write:

$$\varepsilon^{k,\mu+\frac{1}{2}} = \sum_{i=1}^{3} \varepsilon_i^{k,\mu+\frac{1}{2}} \ , \ \varepsilon_i^{k,\mu+\frac{1}{2}} \equiv (T^{k,1})^m \varepsilon_i^{k,\mu}, \quad i = 1,2;$$

$$\varepsilon_3^{k,\mu+\frac{1}{2}} \equiv \{(T^{k,1} + T^{k,2})^m - (T^{k,1})^m\}\varepsilon^{k,\mu}.$$

Choose $\alpha = 4^{-k}/B_3$, then $||\varepsilon_1^{k,\mu+\frac{1}{2}}||_{0,k} \le ||\varepsilon^{k,\mu}||_{0,k}$ and $||\varepsilon_2^{k,\mu+\frac{1}{2}}||_{0,k} \le$ $\le (1-\gamma)^m||\varepsilon_2^{k,\mu}||_{0,k}$. Furthermore, $||T^{k,1}||_{0,k} < 1$, hence $||(T^{k,1}+T^{k,2})^m - (T^{k,1})^m||_{0,k} \le (1+||T^{k,2}||_{0,k})^m - 1 \le \exp(m||T^{k,2}||_{0,k}) - 1$ and $||T^{k,2}||_{0,k} \le 2^{-k}B_4/B_3$, so that

$$||\varepsilon_3^{k,\mu+\frac{1}{2}}||_{0,k} \le \{\exp(2^{-k}mB_4/B_3)-1\}||\varepsilon^{k,\mu}||_{0,k}.$$

Hence, assumption 3 is satisfied with $\theta_\gamma = (1-\gamma)^m$ and $B_6 = mB_4/B_3$. Next, assumption 5 will be verified. Choose m and γ such that $B_3(\gamma+\theta_\gamma) \le (10B_5)^{-1}$, for example, let $\gamma,\theta_\gamma \le (20B_3B_5)^{-1}$, hence m should satisfy, since $\theta_\gamma = (1-\gamma)^m$,

$$m \ge -\ln(20B_3B_5)/\ln(1-\gamma). \tag{5.17}$$

Furthermore, let p be the smallest integer such that

$\exp(B_6.2^{-p-1})-1 \ + B_4.2^{-p-1} \ \le (10B_5)^{-1}$. Then

$$\zeta_k \le \frac{1}{5} \ , \ \ell \ge k \ge p+1 \ , \forall \ \ell. \tag{5.18}$$

We redefine j as follows: j:=max(j,p) (j was previously defined after eq. (5.14)). We have:

$$\eta_k \le B_7^{-1}B_8(\exp(B_6.2^{-k})+\tfrac{1}{3}) = 2\exp(B_6.2^{-k})+2/3,$$

taking (4.21) into account. Let p be the smallest integer such that $\eta_{p+1} \le 3$. Choose $\sigma = 3$. (Smaller values of σ would also do, but it turns out that $\sigma \ge 4$ would increase the asymptotic computational complexity, cf. [4] p.127).
We redefine:

j:=max(j,p)

Then $(\zeta_{j+1}+\eta_{j+1}\zeta_{j+1}^n)^3 \le \frac{1}{5}\ \zeta_j^2(1+3/5)^3 \le \zeta_j^2$. Hence assumption 5 is satisfied.

Theorem 3 gives: $||\varepsilon^{\ell,\mu+3}||_{0,\ell} \leq \frac{1}{25}||\varepsilon^{\ell,\mu}||_{0,\ell}$, and we can now estimate the computational complexity. Of great importance for the computational efficiency is the fact, that all matrices A^k as defined by (4.22), with r^k and p^k given by (4.20) and A^ℓ given by (4.2) are sparse and in fact have the structure of a 9-point finite difference operator (see for example [9] or [10]), so that the multiplication $A^k u^k$ takes only 9.4^k operations. With other prolongation and restriction operators p^k and r^k the density of A^k may increase as k decreases, and the asymptotic computational complexity becomes larger than our estimate below.

First the computational work W_k needed to obtain $u^{k,\mu+1}$ from $u^{k,\mu}$ in algorithm 2 is estimated. Computation of $u^{k,\mu+\frac{1}{2}}$ by one application of M^k takes $O(4^k)$ operations, taking into account the sparsity of A^k as discussed above, and the fact that $m = O(1)$ according to (5.17). Compution of f^{k-1} also takes $O(4^k)$ operations, again taking into account the sparsity of A^k. The coarse grid correction $p^k v^{k-1}$ is obtained at a cost of $3W_{k-1}$ operations, since $\sigma = 3$. Hence

$$W_k = 3W_{k-1} + O(4^k). \tag{5.19}$$

For the purpose of the present asymptotic analysis the $O(4^k)$ term may be replaced by $a.4^k$ with a some constant. In that case the solution of (5.19) is found to be:

$$W_k = W_j 3^{k-j} + \{1-(3/4)^{k-j}\}a.4^{k+1}. \tag{5.20}$$

with W_j the work required so solve for v_j according to algorithm 2.
Hence $W_\ell = O(4^\ell)$ with W_ℓ the work needed for one execution of algorithm 2. It seems reasonable to terminate the iterations when the residue has been reduced by a factor $O(4^{-\ell})$, so that the remaining error is of the same order as the discretization error of the finite difference scheme (4.2). Since $\sigma = 3$ iterations are needed to reduce the residue by a factor 1/25 according to theorem 3.2, we need to execute $O(\ell)$ iterations. Our final conclusion is, that the asymptotic complexity of the multiple grid method considered is at most $O(N \ln N)$, with $N = (2^\ell-1)^2$ the number of unknowns.

This estimate is asymptotically very sharp; only the factor ln N might perhaps be improved upon. But the estimates obtained turn out to be very pessimistic in practice For example, for Laplace's equation one finds $B_3 = 8$, $B_{12} = 2\sqrt{2}$, $B_{14} = 1$, $B_{3j} = 1/6$, $B_{23} \cong 73$, $B_5 \cong 9.85 * 10^4$, $\gamma \cong 0.635 * 10^{-6}$, $m \gtrsim 2.24 * 10^7$. According to the foregoing, the residue will be reduced by a factor of 1/25 by one multiple grid iteration which entails, among other things, m relaxation iterations on the finest grid. In practice the efficiency is found to be much better. For example, Brandt [6] (p.351) reports a total amount of work equivalent to less than 3 fine grid relaxation iterations for a residue reduction by a factor e.

6. FINAL REMARKS

We briefly summarize some results concerning the asymptotic computational complexi-

of the multiple grid approach that have appeared in the literature. Algorithm 2 was first studied by Fedorenko and Bakhvalov [7,8,4]. For A^k, $k < \ell$ they took the same finite difference operator as A^ℓ, with larger step-size. Their p^k and r^k correspond to cubic interpolation and canonical injection (i.e. $u_{ij}^{k-1} = u_{2i,2j}^k$). For the Poisson equation on a rectangle Fedorenko [8] has shown, that the computational complexity is O(N ln N). This was generalized to equation (1.1) on a rectangle by Bakhvalov [4]. A method similar to algorithm 2 with p^k, r^k and A^k as in sections 4 and 5 of the present paper has been considered by Frederickson [9], who proves O(N ln N) complexity for the Poisson equation on a rectangle. The present paper generalizes this to equation (1.1) on a rectangle. Hackbusch [11] has recently given a further generalization to arbitrary regions. For applications of the multiple grid approach to finite element equations, see [1,12].

Numerical experiments indicate, that the efficiency of multiple grid methods does not deteriorate in applications to problems much more general than those considered in the publications just mentioned. See for example [9,10] where the finite difference method is used for the Poisson equation in non-rectangular regions, or [6,15] where succesfull applications are reported to the Navier-Stokes equations and the transonic small-disturbance equation. It is found experimentally, that also in these more complicated applications the computational complexity is roughly O(N).

Acknowledgement. The author is indebted to prof.dr. W. Wetterling and to dr C. Cuvelier for very helpful discussions.

References

1. G.P. Astrakhantsev: An iterative method of solving elliptic net problems. USSR Comp. Math. Math. Phys. 11, no.2, 171-182, 1971.

2. I. Babuška: Error bounds for finite element method. Num. Math. 16, 322-333 (1971).

3. I. Babuška and A.K. Aziz: Survey lectures on the mathematical foundations of the finite element method. In: A.K. Aziz (ed.): The mathematical foundations of the finite element method with applications to partial differential equations. Academic Press, New York and London, 1972.

4. N.S. Bakhvalov: On the convergence of a relaxation method with natural constraints on the elliptic operator. USSR Comp. Math. Math. Phys. 6 no. 5, 101-135 (1966).

5. A. Brandt: Multi-level adaptive technique (MLAT) for fast numerical solution to boundary-value problems. Proc. 3rd Internat. Conf. on Numerical Methods in Fluid Mech. (Paris, 1972), Lecture Notes in Physics 180, 82-89. Springer-Verlag, Berlin and New York, 1972.

6. Brandt: Multi-level adaptive solutions to boundary-value problems. Math. Comp. 31, 333-390 (1977).

7. R.P. Fedorenko: A relaxation method for solving elliptic difference equations. USSR Comp. Math. Math. Phys. 1, 1092-1096 (1962).

8. R.P. Fedorenko: The speed of convergence of one iterative process. USSR Comp. Math. Math. Phys. 4 no. 3, 227-235 (1964).

9. P.O. Frederickson: Fast approximate inversion of large sparse linear systems. Mathematics Report 7-75, 1975, Lakehead University.

10. W. Hackbusch: On the multi-grid method applied to difference equations. Computing 20, 291-306 (1978).

11. W. Hackbusch: Convergence of multi-grid iterations applied to difference equations. Math. Inst., Universität zu Köln, Report 79-5, April 1979.

12. R.A. Nicolaides: On multiple grid and related techniques for solving discrete elliptic systems. J. Comp. Phys. 19, 418-431 (1975).

13. R.A. Nicolaides: On the ℓ^2 convergence of an algorithm for solving finite element equations. Math. Comp. 31, 892-906 (1977).

14. J. Nitsche and J.C.C. Nitsche: Error estimates for the numerical solution of elliptic differential equations. Arch. Rat. Mech. Anal. 5, 293-306 (1960).

15. P. Wesseling: Numerical solution of the stationary Navier-Stokes equations by means of a multiple grid method and Newton iteration. Report NA-18, Delft University of Technology, 1977.

16. P. Wesseling: A convergence proof for a multiple grid method. Report NA-21, Delft University of Technology, 1978.